Third Edition

DRUGS AND ALCOHOL

Kenneth L. Jones
Louis W. Shainberg
Curtis O. Byer

MT. SAN ANTONIO COLLEGE

HARPER & ROW, PUBLISHERS

New York   Hagerstown   San Francisco   London

COVER PHOTO: brown and white capsule: Phenaphen containing phenobarbital (barbiturate); pink and white capsules: Benadryl (amphetamine); white tablet: Tylenol with codeine (synthetic narcotic); black and green capsule: Librium (tranquilizer). Alcohol: Rye whiskey.

Photo by Beckwith Studios.

Sponsoring Editor: Kyle Wallace
Project Editor: Karla B. Philip
Designer: Helen Iranyi
Production Supervisor: Stefania J. Taflinska
Compositor: Maryland Linotype Composition Co., Inc.
Printer and Binder: The Murray Printing Company
Art Studio: Danmark & Michaels, Inc.

DRUGS AND ALCOHOL, Third Edition

Copyright © 1979 by Kenneth L. Jones, Louis W. Shainberg, and Curtis O. Byer.

All rights reserved. Printed in the United States of America. No part of this book may be used or reproduced in any manner whatsoever without written permission except in the case of brief quotations embodied in critical articles and reviews. For information address Harper & Row, Publishers, Inc., 10 East 53rd Street, New York, N.Y. 10022.

Library of Congress Cataloging in Publication Data

Jones, Kenneth Lamar, Date —
  Drugs and alcohol.

  Bibliography: p.
  Includes index.
  1. Narcotic habit. 2. Alcoholism.
I. Shainberg, Louis W., joint author. II. Byer, Curtis O., joint author. III. Title.
RC566.J63   1979   613.8′3   78-15391
ISBN 0-06-043436-8

CONTENTS

Preface viii

chapter 1 PROPER USE OF DRUGS 1

A Drug 1
Medical Use of Drugs 2
Administration of Drugs 2
Treatment of Diseases and Conditions 4
Names of Drugs 5
Use of Drugs 6
Major Actions and Effects of Drugs 20
Medical Classification of
   Drug Actions and Effects 21
Summary 24

chapter 2 NEUROLOGICAL ASPECTS OF ABUSIBLE DRUGS 28

How Psychoactive Drugs Work 30
Drug-Induced Effects on the
   Central Nervous System 33
Structural Components of the
   Central Nervous System 34

CONTENTS

Functional Components of the
   Central Nervous System   39
Spectrum and Continuum of Drug Actions   47
Summary   52

chapter 3   DRUGS COMMONLY ABUSED   56

Anesthetics   57
Narcotics   59
Volatile Liquids   70
Hypnotic-Sedatives   73
Antihistamines   79
Tranquilizers   82
Cannabis (Marijuana)   84
Hallucinogens   90
Cocaine   96
Antidepressants   98
Amphetamines   101
Poly Drug Use   106
Summary   107

chapter 4   PSYCHOLOGICAL AND SOCIOLOGICAL FACTORS IN DRUG ABUSE   115

Patterns of Drug Abuse   115
Psychological Factors in Drug Abuse   122
Sociological Factors in Drug Abuse   129
Interactions of Factors   133
Summary   133

chapter 5   ENFORCEMENT AND/OR TREATMENT?   136

Philosophies of Drug-Abuse Control   137
Legal Controls of Drug Abuse   137
Drug Control and Law Enforcement   146
Drug-Abuse Treatment   148
Return to Society   156
Toward the Future   157
Summary   157

DRUGS AND ALCOHOL

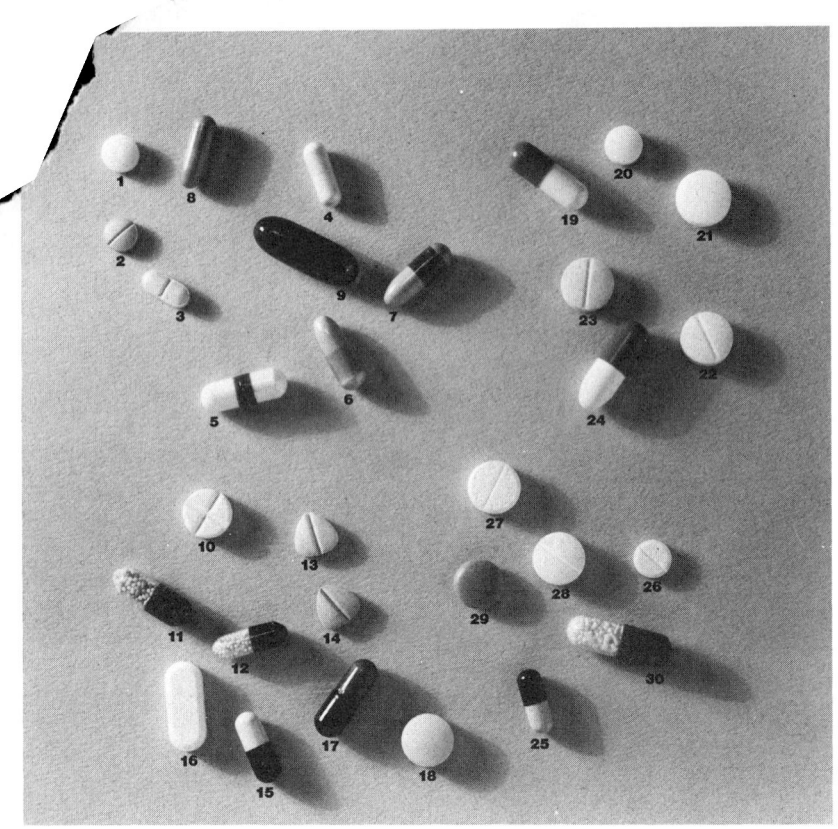

DRUGS COMMONLY ABUSED. The four groups of drugs in this picture show many of the drugs with potential for being abused; they may produce personality changes, euphoria, or abnormal social behavior. The drugs have been numbered to allow reference to the color index below.

Barbiturates (hypnotic–sedatives): 1. (white tablet) Phenobarbital 1 gram **2.** (blue-green tablet) Butisol ½ gram **3.** (pink tablet) Amytal 1½ grams **4.** (yellow capsule) Nembutal 1½ grams **5.** (white/blue/white capsule) Carbrital **6.** (blue capsule) Sodium Amytal 3 grams **7.** (blue/red capsule) Tuinal 3 grams **8.** (red capsule) Seconal 1½ grams **9.** (green capsule) Chloral Hydrate 7½ grams

Amphetamines: 10. (round white tablet) Methadrine 10 milligrams ("speed") **11.** (green capsule) Dexamyl No. 2 **12.** (brown/orange capsule) Dexedrine 15 milligrams **13.** (orange heart-shaped tablet) Dexedrine **14.** (blue heart-shaped tablet) Dexamyl **15.** (white/black capsule) Diphetamine 12½ milligrams **16.** (long white tablet) Tenuate Dospan **17.** (red/maroon capsule) Biphetamine T-20 **18.** (pink tablet) Preludin Endurets

Synthetic Narcotics: 19. (blue/yellow capsule) Fiorinal with Codeine ½ gram **20.** (small white tablet) Demerol 100 milligrams **21.** (big white tablet) Empirin Compound with Codeine ½ gram **22.** (yellow tablet) Percodan **23.** (pink tablet) Demi-Percodan **24.** (white/gray capsule) A.S.A. with Codeine ½ gram

Tranquilizers: 25. (green/black capsule) Librium 10 milligrams (minor tranquilizer) **26.** (yellow tablet) Valium 5 milligrams (minor tranquilizer) **27.** (white tablet) Equanil 400 milligrams (minor tranquilizer) **28.** (white tablet) Miltown 400 milligrams (minor tranquilizer) **29.** (orange tablet) Thorazine 50 milligrams (major tranquilizer) **30.** (blue/yellow capsule) Meprospan-400 (minor tranquilizer)

CONTENTS

chapter 6 ALCOHOLIC BEVERAGES: USE AND ABUSE 162

    A Socially Acceptable Mood-Modifying Drug   163
    Alcoholic Beverages   163
    Effects of Alcohol on the Human Body   166
    Proper Use of Alcoholic Beverages   175
    Young People and Alcohol   177
    Problems Resulting From Alcohol Abuse   177
    Summary   180

chapter 7 ALCOHOLISM (ALCOHOL DEPENDENCE) 183

    Meanings of "Alcoholism"   184
    Causes of Alcoholism   185
    Alcoholism and Women   186
    Older Problem Drinkers   187
    Phases of Alcoholism   187
    Treatment of Alcoholism   193
    Summary   197

Glossary   199

Bibliography   207

Index   211

PREFACE

We live in a society in which most people regularly use drugs or substances containing drugs for medication or recreation. There are also some individuals who, for a variety of reasons, *abuse* drugs, thereby creating problems for themselves as well as for society. Some drugs, such as those that combat disease, are indispensable to society. Other recreational drug-containing substances—coffee (containing caffeine), tobacco (nicotine), and alcoholic beverages—while not essential to society, are accepted by the majority of people.

Each of us must make decisions regarding the use of drugs and alcohol. We must also act as informed members of society in voting or otherwise expressing our opinions concerning the legality of drugs and drug-containing substances used for recreational purposes. Such important decisions should be based on facts, not just emotions. It is the purpose of *Drugs and Alcohol,* Third Edition, to provide you with these facts; you must make the decisions.

We have found that most books on drugs or alcohol present some of the pertinent facts, speculate on a few more, and neglect a great many. In an effort to correct this situation, *Drugs and Alcohol* objectively presents, without speculation, the basic information vital in making important decisions concerning the medical and recreational use of drugs and alcohol.

Drugs and Alcohol is an outgrowth of common problems and anxieties associated with drug and alcohol use. Its contents have resulted from many years of communication with young people, both inside and outside the college classroom. We have tried to answer, as completely as possible, the questions most often asked by these students and have endeavored to keep the discussion open, honest, and up to date, both in factual content and in philosophy.

The laws and restrictions governing drug distribution are discussed at length. The medical use of drugs by the professional (physician, dentist) and by the individual (self-medication) is also discussed. The physiological effects of drugs and the most susceptible parts of the central nervous system are also investigated. The placing of drug groups within a *continuum of actions and effects* on the central nervous system was initially presented in the first edition. This classification of drugs was well received, illiciting favorable response from both students and teachers, as well as from professionals working in the field of drug abuse. This approach to discussing drug effects has been further refined in this third edition.

Because the use, and particularly the abuse, of alcohol has become a growing problem in our society, we have addressed this problem by including material on the young problem drinker, as well as alcoholism in women and the elderly. We have also endeavored to look into the causes of alcoholism in more detail. The physical, psychological, and medical aspects of alcoholism are presented in depth, and the recreational use and abuse of alcohol is discussed to assist an individual in recognizing a problem in themselves as well as in others.

The authors are indebted to Robert W. Earle, Ph.D., Senior Lecturer at the Department of Medical Pharmacology and Therapeutics, University of California, Irvine, for his assistance in developing the Continuum of Drug Actions and Effects. We would also like to acknowledge the help of Detective Larry Zambrano, Police Department, Pomona, California, for his review of the drug laws.

Kenneth L. Jones
Louis W. Shainberg
Curtis O. Byer

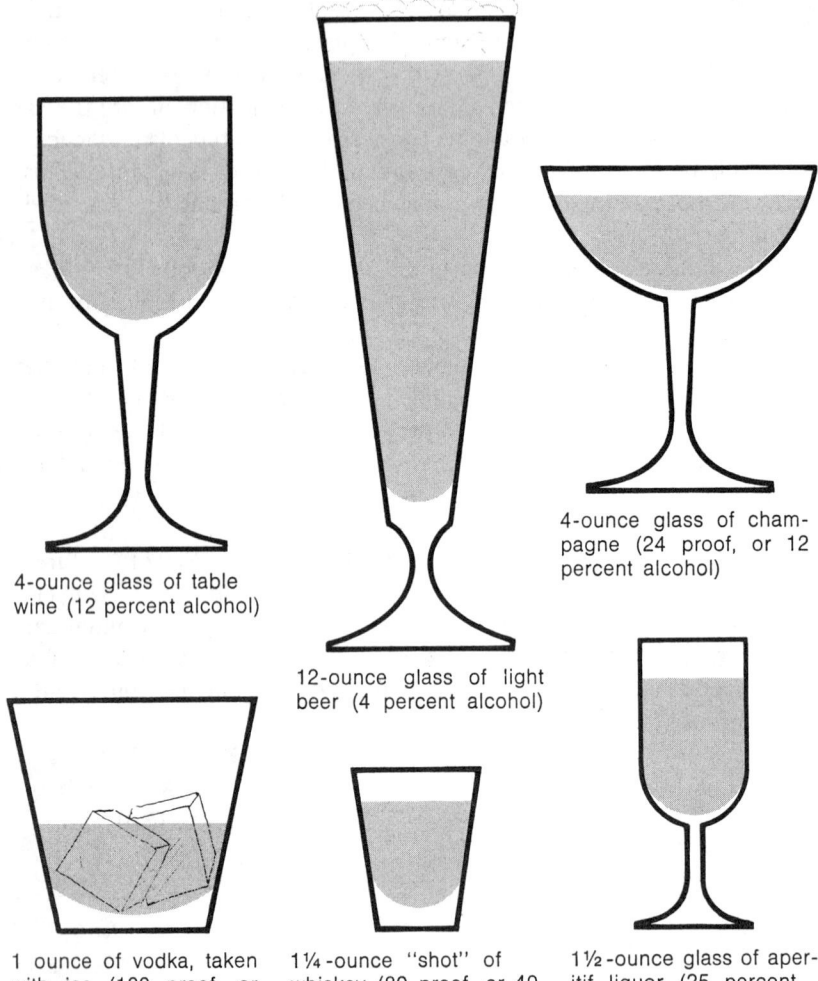

4-ounce glass of table wine (12 percent alcohol)

12-ounce glass of light beer (4 percent alcohol)

4-ounce glass of champagne (24 proof, or 12 percent alcohol)

1 ounce of vodka, taken with ice (100 proof, or 50 percent alcohol)

1¼-ounce "shot" of whiskey (80 proof, or 40 percent alcohol)

1½-ounce glass of aperitif liquor (25 percent alcohol)

Equivalent Quantities of Alcohol. Each of these "social drinks," containing roughly equivalent amounts of alcohol, would produce an average blood alcohol level of 0.03 percent in most individuals. Consumption of one of these drinks may create no apparent change in physical coordination; yet alcohol, affecting the brain, becomes a factor in traffic accidents at blood alcohol levels even as low as 0.03 percent. The average person can oxidize, or eliminate from his system, only one of these drinks during a period of 1 hour.

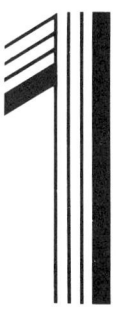

PROPER USE OF DRUGS

A DRUG

In contemporary society the word *drug* has two connotations—one positive, explaining its crucial role in medicine, and one negative, reflecting, not the natural and synthetic makeup of these chemicals, but the self-destruction and socially deleterious patterns of misuse. To the general public, a drug is a medicine, a substance used in the treatment of a disease. However, a more complete or scientific definition for a drug is: any substance, other than food, which when introduced into the body, alters the body or its functions. The public is familiar with many drugs and their uses, for example, the antibiotics used in the treatment of bacterial infections and the narcotics used in relieving pain. The value of such drugs has been established through years of research, evaluation, and use.

It is equally well known that the actions and effects of all drugs are not alike. Some are rather mild in effect and relatively harmless; others are severe in effect and can be extremely dangerous if not taken exactly as prescribed by a physician. Some drugs, such as LSD, are considered so dangerous that laws have been passed to prohibit their use except in research projects. Other drugs, such as marijuana, have little valid medical use and are sufficiently dangerous when misused to be prohibited legally. No drug is considered safe for the public to use without a prescription if its effects are dangerous enough to create a potential hazard to the person using it.

MEDICAL USE OF DRUGS

Drugs have been used for thousands of years. At first they were used by societies as part of religious healing ceremonies; however, more reliance was placed on prayers, incantations, and charms than on the specific drug used. Eventually, certain drugs proved to be effective, and their powers became highly guarded secrets, known only to the people who governed their use.

It was not until well into the nineteenth century that scientists began to make accurate experiments to discover precisely what chemicals were contained in drugs and what effects individual drugs might have.

In the past 30 years, many new drugs have been introduced to the general public, revolutionizing the practice of medicine, in many ways bypassing natural law, and overcoming death. But like all revolutions, this one carries with it the risk of excesses—abuse.

Because of the development of the so-called "wonder drugs," many people have built up unrealistic expectations about what drugs can do for them. In other words, the people of the United States are, in general, "drug users." In their zealous search for miraculous cures, they often decide for themselves what drugs and what quantities of drugs are needed instead of leaving this difficult and delicate decision where it belongs—in the hands of trained physicians and dentists. The authors of this book feel that some understanding of how drugs act and of what can be expected of them may help put the myth of miracle drugs into its proper perspective.

ADMINISTRATION OF DRUGS

Drugs can be administered in a number of ways, and a physician makes the decision on the basis of the effect sought after. Figure 1.1 shows the various ways in which drugs can be introduced into the body. After being taken into the body, drugs may be distributed by the blood stream to the many organs, tissues, and cells.

Drugs can enter the blood stream by being taken *orally* (by mouth) and absorbed from the digestive tract. In general, when a choice of methods is possible, physicians prefer oral administration because this method assures a more gradual and sustained effect on the body, is less painful, is less likely to produce adverse reactions, and is safer than other methods of administration [skin is not punctured or chance of damage to lungs]. Also, oral administration is convenient for the patient; the drugs can be taken at home without the need for the presence of a physician or a nurse.

When a physician prescribes medicines to be taken orally, the time at which they should be taken is usually specified, depending upon the

ADMINISTRATION OF DRUGS

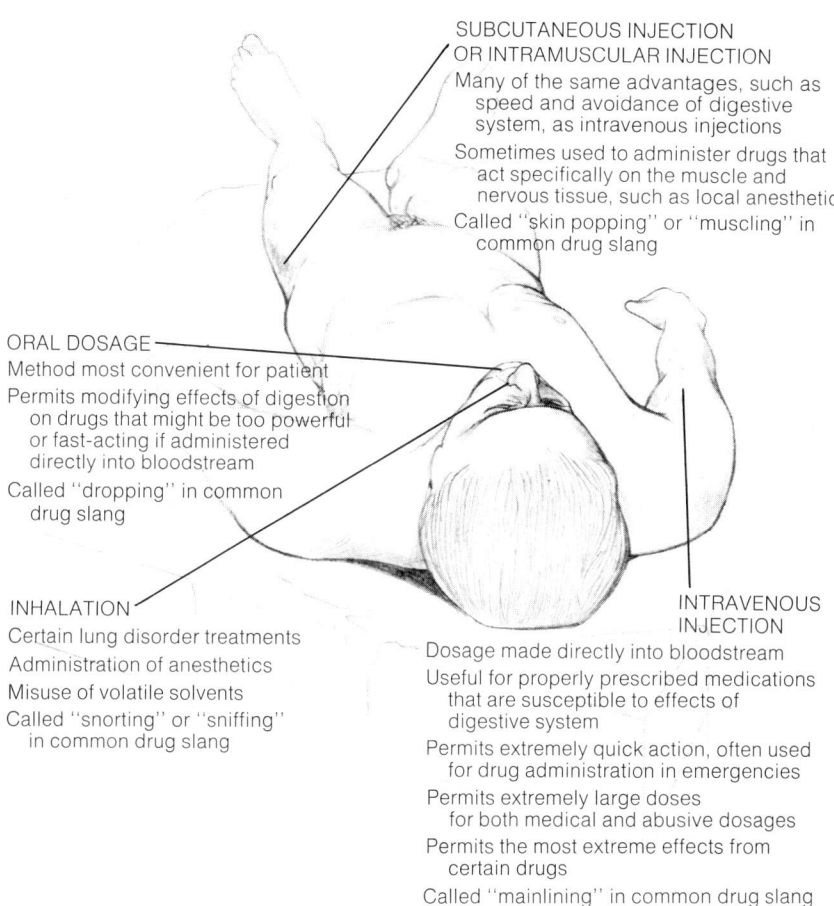

Figure 1.1. *Routes of administration of drugs.*

kind of action desired. If rapid action is needed, the individual is instructed to take the drug before meals, so that the digestion of food will not interfere with the absorption of the drugs into the blood stream. If the medicines may irritate the stomach or if a more sustained action is needed, the individual may be instructed to take the drug after meals. Drugs that produce sleep or a sedative effect are, of course, taken at bedtime.

Drugs also may be *inhaled* into the lungs and enter the blood stream from there. Inhalation may be used in the treatment of certain lung disorders or in the administration of an anesthetic.

Certain drugs must be administered by *injection* (under the skin, into a muscle, or into a vein). Some medicines cannot be absorbed into the blood stream through the digestive tract; some are irritating to it or to the lungs; and some drugs are destroyed by the gastric acid of the stomach. Such drugs are usually injected into the body.

TREATMENT OF DISEASES AND CONDITIONS

Often people seeking relief from an illness are impatient and resentful toward a physician who does not seem to be doing anything to effect a cure. They may feel that an office call is a waste of time unless the physician hands them a prescription to be filled. However, in many cases the prescribing of drugs is exactly what should *not* be done. A physician does not prescribe until it is known what is being treated. The process of making a diagnosis may take time. If drugs are prescribed too quickly, they may temporarily relieve the symptoms and mislead both the patient and the physician into thinking the illness has been cured. Or, the drugs may interfere with the diagnostic procedures used by the physician to find out the exact illness or its cause.

Only after a diagnosis has been made can a physician prescribe an appropriate form of treatment. Sometimes drugs are not prescribed because the disorder will correct itself without treatment or because another form of treatment would be more effective. In some cases nothing other than reassurance is needed.

Types of Therapy

Treatment, or *therapy*, may be defined as any effort to cure a disease or condition, arrest its course, lessen its severity, or alleviate the pain and inconvenience caused. There are many different types.

Dietary therapy. Dietary therapy is the treatment of disease by regulation of the food the patient eats. In the treatment of all conditions, it is important that a patient have proper food and a proper amount of fluid.

Operative therapy. Operative therapy, or surgical treatment, is a mechanical treatment of disease. The diseased organ or area is removed, repaired, opened, drained, or manipulated by the surgeon.

Physical therapy. Physical therapy is the use of physical agents such as heat, cold, light, radiation (X-rays, infrared, and ultraviolet rays), electricity, and massage in the treatment of a condition.

Psychotherapy. Psychotherapy is the treatment of emotional disorders. Its purpose is to remove or reduce the emotional factors in disease, to dispel the worries, doubts, and fears that are so often the cause of illness and that nearly always affect organic diseases.

NAMES OF DRUGS

Chemotherapy. Chemotherapy is the use of drugs to treat disease. Any chemical or mixture administered to a sick person is a drug or medicine. These are derived from various natural sources or are synthetically manufactured in the laboratory.

What a Patient Can Do

The benefits to be derived from drugs prescribed by a physician and taken in accordance with instructions far outweigh the possible risks involved in their use. Since the chemistry of the body is subtle and variable, only a physician should have the responsibility of prescribing and directing the use of drugs in the treatment of illnesses.

Individuals can help themselves and assist their physicians by taking drugs only in the amount and at the times instructed; by reporting to their physicians any adverse reaction they have to a drug; by being as skeptical as their physicians are about new drugs until they have been thoroughly tested; by not treating their own illnesses on the basis of what they read, see on television, or hear from the neighbors; and by not taking drugs that have been prescribed for others.

NAMES OF DRUGS

Many drugs have been identified by two or more names. The *official name* is the name listed in *The Pharmacopeia of the United States*. The *chemical name* (used mainly by chemists) is a very precise description of the chemical makeup of the drug. It shows the exact placement of atoms or atomic groups. For example, the chemical name of the antibiotic *tetracycline* is: 4-dimethylamino-1,4,4a,5,5a,6,11,12a-octahydro-3,6,10,12, 12a-pentahydroxy-6-methyl-1,11-dioxo-2-naphthacenecarboxamide.

Before a drug may be sold, it is assigned a *generic name* by the American Medical Association Council on Drugs and the World Health Organization. The generic name is never changed and is usually the same in all countries. This name is simpler than the chemical name, yet often reflects the chemical family to which the drug belongs. The name *tetracycline* is a generic name. The generic name is never capitalized.

The *trademark name* or *brand name* usually appears with the sign ® at the upper right of the name (indicating that the *name* is registered and that its use is restricted to the manufacturer who is the legal owner of the *name*). The first letter of the name is capitalized. For example, tetracycline is known under the brand names Achromycin, Panmycin, Polycycline, Tetracyn, and Tetracyn V. Sometimes a single drug may be sold under 10 or 20 different brand names. Much controversy exists regarding whether drugs should be prescribed by brand names or the often lower-priced generic name. Further brand name confusion is created when mixtures or combinations of drugs are sold under brand names that do not reflect their contents. For example, mixtures of phenacetin, aspirin, and caffeine are sold under brand names of Emperin, APC, PAC, as well as a number of brand name aspirin compounds.

The word *proprietary* is used when referring to drugs or names of drugs that belong to specific manufacturers.

Common, street, or *slang* names, terms, and expressions are the language of illegal drug use and abuse. There are variations from one geographic area to another, and the street language is constantly changing. Often, drug abuse language is picked up and incorporated into contemporary slang by nonabusers, particularly teenagers and young adults. Consequently, use of many of these terms cannot be considered evidence of drug abuse.

USE OF DRUGS

If drugs are prescribed too quickly, they may temporarily relieve the symptoms and mislead both patient and physician into thinking the illness has been cured, or they may interfere with the diagnostic procedures being used by the physician. Sometimes the physician does not prescribe drugs because the disorder will correct itself without treatment or because another form of treatment would be more effective. Some physicians do prescribe drugs more readily than others, and some patients take prescribed medicines more readily than others.

Prescription and Nonprescription Drugs

Before the twentieth century, illness was treated with well-known home remedies often made from local plants, time-honored preparations, or substances of "secret" composition sold to the public. The twentieth century ushered in an era of rapidly expanding drug development, and it soon became evident that federal legislation was needed to protect the public in areas such as drug use where it was unable to protect itself. Between 1906 and 1938 drugs had become so numerous and complex that federal laws that restricted the public sale of drugs were passed. These laws required that specific drugs only be dispensed by qualified pharmacists under the directions (prescription) of a physician, dentist, or veterinarian.

In 1952 the Durham-Humphrey amendment of the 1938 Federal Food, Drug, and Cosmetic Act specifically divided drugs into: (1) those drugs safe for use without medical supervision, which therefore may be sold over the counter and (2) those drugs not safe for unsupervised use, which therefore are required to bear a label with the legend: "Caution: Federal law prohibits dispensing without prescription." Every year there are further modifications affecting prescription and nonprescription (over-the-counter) drugs.

Self-Medication

In the United States nearly $2 billion are spent yearly for over-the-counter drugs that do not require a prescription or the advice of a physician. These drugs can induce sleep or wakefulness, relieve pain or tension, or supply the body with vitamins and minerals. Substances that affect any part of the body from head to toe can be purchased. Such widespread use of self-medication is primarily the result of advertising through mass communication media (magazines, newspapers, books, radio, and especially television). Although the products are commonly called *patent medicines,* a more correct name is *proprietary* compounds, since the formulas are seldom patentable. Most patent medicines are of little value but usually do not cause any direct harm. However, some do have potentially harmful effects if used to excess or by a physiologically sensitive person. Certain medicines that are normally harmless can become very dangerous when used in combination with a particular physical condition. For example, usually harmless laxatives may be quite hazardous when taken by a person with appendicitis. Other products may produce so much drowsiness that driving a motor vehicle becomes hazardous.

Probably the most effective substance sold without a prescription is aspirin (acetylsalicylic acid), which today forms the basis for a myriad of proprietary compounds. The beneficial properties of aspirin, in addition to relief of pain, are the reduction of fever and inflammation. The "glorified aspirin" products—which usually contain aspirin, phenacetin, and caffeine, or just aspirin and caffeine—along with the buffered aspirin products, have been shown to be no more effective for most individuals than plain aspirin. Several products that originally contained phenacetin no longer do, because (in large doses) this agent has been shown to cause permanent kidney damage. Aspirin should be purchased in small quantities (about a three-month supply), because it breaks down in storage. The breakdown is indicated by a vinegary odor in the bottle, owing to the presence of acetic acid.

Of the more than $2 billion spent in the United States each year on nonprescription remedies about $350 million goes for aspirin and its products and another $350 million goes for cold remedies. Every year new "miracle" cold remedies are offered to the public with great fanfare from their manufacturers, only to drop quietly out of the picture a few years later, when the producer releases a newer "miracle." The fact remains that, despite the many advances in other areas of medicine, there is still no way to prevent or cure the common cold. Some of the products offered do provide some symptomatic relief, such as opening a stuffed nose, drying up nasal drip, and reducing fever, but the best advice for the early cold sufferer still is to go to bed for 24 hours. This rest helps prevent secondary infections and complications. The actual cold will last no longer than a week, but these secondary problems may last considerably longer. Antibiotics have no effect upon the actual cold, because antibiotics are not effective against viruses. Antibiotics should be reserved only for those cases in which secondary bacterial infections occur. The cold sufferer should not press a physician into prescribing antibiotics.

Self-medication will always be controversial, with adherents for and against it. The public is health conscious, and many human ailments are trivial and temporary. People seek to alleviate these discomforts as quickly and as economically as possible. Although self-medication may harm an individual, the hazards can be minimized by educating the public about drugs.

Today drugs can be bought in supermarkets, restaurants, and from vending machines. Sales promotion on radio and television encourages self-medication for real or imagined illnesses. But remember, overuse or abuse of any drug can cause detrimental effects. As shown in Table 1.1, drugs considered harmless by many, such as aspirin or vitamins, when taken regularly and in large amounts can produce untoward effects. Aspirin can cause depression of normal bone marrow function, causing severe anemia. Multiple vitamin compounds that contain groups of minerals can cause stomach and intestinal disturbances. And vitamins, such as C and D, in excessive amounts in time can cause permanent kidney damage. Continued self-medication of a recurring ailment may mask a serious condition, endangering life and creating a need for prolonged and expensive medical treatment.

Directions for Use of Nonprescription Drugs

Many over-the-counter preparations are limited in their effectiveness. But a nonprescription drug, like a prescription drug, must be proved safe and relatively effective for treatment of the conditions it is sold to remedy. Nonprescription drugs pose no threat to the average person when *used as directed*. The key to the use of nonprescription drugs is reading and *understanding* the label. A typical label is shown in Figure 1.2.

Most over-the-counter products contain some warning regarding their use. These warnings include: avoidance of use by children, avoidance of chronic use, and avoidance of use during pregnancy and in the presence of specific diseases or conditions. Typical warnings to look for include:

1. How to use medication safely.
2. When not to use the preparation (see Table 1.1).
3. When to stop taking the drug.
4. When to see a physician.

TABLE 1.1
Drug Interactions[a]

Primary Drug	Drugs that Interact with Primary Drug	Reactions that May Be Produced and Comments
Alcohol, central nervous system (CNS) depressant, inhibits some enzymes and enhances others.	Anesthetics, general (used in surgery)	Heavy, problem, or alcoholic drinkers who have produced an alcohol *tolerance* require a larger amount of anesthetics.
	Anticoagulants (used to stop formation of blood clots). Examples: Coumadin, Dicumarol, Heparin, Panwarfin, etc.	Alcohol inhibits actions of anticoagulants, but response is unpredictable and variable. Individuals using these drugs should not drink alcoholic beverages.
	Antidepressants (tricyclics). See antidepressants under Primary Drug heading for examples.	Can be a lethal combination. Potentiate sedative actions with alcohol.
	Antidiabetic drugs (oral forms). Examples: DBI, Diabinese, Dymelor, Orinase, etc.	Alcohol increases metabolism of oral antidiabetics; can decrease time of effectiveness by 50%. Alcohol may produce low-blood-sugar convulsions, especially in children.
	Aspirin (Salicyclates): See Aspirin under Primary Drug heading for examples.	Aspirin-containing medications may cause bleeding of the stomach and intestinal walls (ulcers). Irritation and bleeding actions intensified by alcohol.
	Central Nervous System (CNS) depressants. Examples: hypnotics, narcotics, tranquilizers (see sections in chapter for specific examples).	Potentiate alcohol increasing the risk of death by impairing psychomotor skills. Extreme overdose results in death from respiratory failure.
	Central Nervous System (CNS) stimulants. Examples: amphetamines, caffeine, Preludin,	Antagonize the sedative effects of alcohol except that they do not improve the decreased motor function induced by alcohol.

TABLE 1.1 *(Continued)*

Primary Drug	Drugs that Interact with Primary Drug	Reactions that May Be Produced and Comments
	(see sections in chapter for specific examples).	
	Disulfiram (used in treatment of alcoholism), Antabuse, TETD, etc.	Inhibits enzyme metabolizing acetladehyde causing buildup in blood. Extreme nausea, vomiting, etc. Should never be given to someone while intoxicated from alcohol; can be lethal.
	Insulin (injectable antidiabetic)	Alcohol is hypoglycemic, potentiates insulin. May induce extreme low blood sugar resulting in irreversible brain damage, coma, and death.
	Nitroglycerin (used in treatment of cardiovascular disease). Examples: Amyl nitrite, Nitranitol, Peritrate, Vasodiatol, etc.	Severe low blood pressure due to additive vasodilator effect with alcohol. May cause cardiovascular collapse and death.
	Sulfonamides (used in antibacterial treatment)	Inhibits metabolism of acetaldehyde, causing nausea, vomiting (see Disulfiram).
	Vitamin B_{12}	Alcohol-caused malabsorption of vitamin B_{12}.
Amphetamines (see section in chapter for actions and examples).	Alcohol	See CNS comments for Alcohol under Primary Drug heading.
	Antihistamines	Often combined to counteract sedative effect. See antihistamines.
	Antacids	Decrease urinary excretion of amphetamines thereby potentiating them.
	Cocaine	Potentiates amphetamines. May cause vasomotor collapse and respiratory arrest. Lethal.
	MOA inhibitors (used in treatment of emotional dis-	Potentiate amphetamines by slowing rate of metabolism. May cause extreme reactions;

USE OF DRUGS

Table 1.1 *(Continued)*

Primary Drug	Drugs that Interact with Primary Drug	Reactions that May Be Produced and Comments
	orders). Examples: Marplan, Niamid, Nardil, etc.	headache, brain hemorrhage, extreme high blood pressure; death may result.
Antacids (used to neutralize stomach). Examples: Alka-Seltzer, Alka-2, Amphojel, Camalox, Maalox, Mylanta, etc.	Antibiotics (used in treatment of bacterial diseases)	Antacids greatly decrease the absorption of oral antibiotics from the intestine, reducing their effectiveness.
	Anticoagulants (see Alcohol)	Antacids inhibit absorption from intestine.
	Digitalis (used in treatment of heart disease)	Antacids decrease or cause digitalis to be ineffective in treatment of heart disease. Antacids should never be taken with digitalis.
	Aspirin (see Aspirin)	Antacids decrease absorption decreasing their effectiveness.
Antibiotics (used in treatment of bacterial and other microbial diseases). Examples: Acromycin, Erythrocin, Erythromycin, Linocomycin, Nilstat, Penicillin, Streptomycin, Terramycin, Tetracycline, etc.	Antacids (see Antacids)	See Antacids.
	Anticoagulants (oral) (see Alcohol)	Antibiotics given in large doses for prolonged periods increase anticoagulant activity. Tetracyclines produce some anticoagulant activity themselves, causing an additive effect.
	Cyclamates (artificial sweetener)	Cyclamates inhibit action of linocomycin by decreasing its absorption.
	Milk	Milk and milk produce inhibit absorption of Tetracyclines.
	Vitamin B$_{12}$	Neomycin inhibits absorption of vitamin B$_{12}$.
	Vitamin K	Antibiotics destroy the bacteria that produce vitamin K in the large intestine. Vitamin K is used in normal clotting; lack of vitamin K reduces blood's ability to clot.
Antidepressants (trycyclics) (used in treatment of	Alcohol	A lethal combination. Potentiates sedative effects of alcohol. Death has occurred.

TABLE 1.1 *(Continued)*

Primary Drug	Drugs that Interact with Primary Drug	Reactions that May Be Produced and Comments
emotional disorders). Elavil, Niamid, Marplan, Tofranil, etc. Psychotoxic in some individuals.	Amphetamines	Increases activity and side effects.
	Aspirin	May be a lethal combination. Death has resulted from both being used on same patient.
	Barbitures	Potentiates sedative effects of barbiturates. A dangerous combination.
	Narcotics	Potentiates depressant effects.
	Tranquilizers (minor)	Increased depressant effect (additive effect).
Antihistamine-containing drugs. Extremely wide range of over-the-counter and prescription drugs. Examples: allegens, antiasthma, antinauseants (air, car, and sea sickness), cough suppressant, decongestants, sleeping aids, tension relievers. Examples: Allerest, Benadryl, Cheracol D, Congespirin, Coricidin, Dimacol, Dormin, Nyquil, Nytol, Sleep-Eze, Sominex, Triaminic, Vistril.	Alcoholic beverages	Both are CNS depressants and potentiate each other. Some antihistamine compounds also contain amphetamine compounds that, when combined with alcohol, may cause dangerous levels of high blood pressure.
	Anticoagulant drugs	Greatly decrease or reverse effects by forming enzymes in the body that destroy anticoagulant drugs.
	CNS depressants	See Alcohol under Primary Drug heading.
Antihypertensive drugs (used in treatment of high blood pressure). Examples: Ansolysen, Apresoline, Capla, Diuril, Ecolid, Eutonyl, Hydromox, Hyperstat, Inversine,	Alcohol	Potentiates some agents by increasing blood pressure.
	Anesthetics	Potentiate effect of such drugs. Severe potentiation can cause shock or cardiovascular collapse. Surgeon must know when patient is taking antihypertensive drugs.

TABLE 1.1 *(Continued)*

Primary Drug	Drugs that Interact with Primary Drug	Reactions that May Be Produced and Comments
Lasix, Priscoline, Regitine, etc.	Antidepressants (see Alcohol)	Antagonize antihypertensive drugs, destroying their effectiveness. Extreme action can cause severe low blood pressure.
	Cocaine	Reverses the action of some antihypertensive drugs.
	Tranquilizers	Valium may potentiate antihypertensive drugs. Causes overmedication.
	MOA inhibitors (see Amphetamines)	Some MOA inhibitors and some antihypertensive drugs mutually lower blood pressure to a dangerous level unless dosages are adjusted.
	Amphetamines (see Amphetamines	Inhibit some antihypertensive drugs, decreasing their effectiveness.
Anti-inflammatory, drugs (Enzyme products such as: Ananase, Buclamase, Chymar, Varidase, etc. Also hormones and other agents such as: aminopyrine, hydrocortisone, pryazolones, salicylates.	Anticoagulants (see alcohol)	Anti-inflammatory drugs should be given very carefully with anticoagulants. Great possibility of extreme reduction in blood-clotting abilities of blood.
	Sedatives and hypnotics (especially phenobarbitol)	Many inhibit anti-inflammatory agents.
Aspirin containing analgesics (pain suppressants) and other aspirin-containing compounds. Examples: Alka-Seltzer, Anacin, Bufferin, Cama, Cirin, Congespirin, Cope, Coricidin, Doan's Pills, Empirin, Excedrin, Fizrin, Florinal, Measurin, Midol,	Alcoholic beverages	See Alcohol under Primary Drug heading.
	Anticoagulant drugs	The major contraindication of the use of aspirin is with anticoagulant drugs. Aspirin increases the anticlotting effects of these drugs to a point where the person is in danger of internal bleeding.
	Antidiabetic drugs (oral and injectable)	Low blood sugar may be further depressed by potentiation with aspirin.

TABLE 1.1 *(Continued)*

Primary Drug	Drugs that Interact with Primary Drug	Reactions that May Be Produced and Comments
Pamprin, Phenaphin, Vanquish, etc.	Hypertensives (drugs controlling high blood pressure)	Many of these medications also contain caffeine, which can reduce or reverse the effectiveness of the hypertensive drugs.
Barbiturates (major sedatives and hypnotics). (See section in text for examples.)	Alcohol	Alcohol and barbitures is potentially lethal. *Dual* potentiation greatly increases the deadly effect by respiratory failure.
	Antacids (see Antacids)	Decreasing absorption of barbiturates severely nullifies the hypnotic effect.
	Androgens (male hormones)	Barbiturates increase their metabolism, inhibiting their effectiveness.
	Anesthetics	Barbiturates potentiate anesthetic action. May cause delayed recovery and in extreme reactions possible collapse.
	Anticoagulants (see Alcohol)	Barbiturates cause production of enzymes that destroy anticoagulants. Inhibition may last up to six weeks after withdrawal of barbiturates. Also inhibit absorption of anticoagulants.
	Antidepressants (see Alcohol)	Potentiate effects of barbiturates. Severe depression may result.
	CNS depressants	See Alcohol under Primary Drug heading.
	Oral contraceptives	Barbiturates increase the rate of metabolism; can cause them to be ineffective.
	Vitamin C (ascorbic acid)	Vitamin C decreases excretion of barbiturates, increasing time of sedation. Barbiturates increase excretion of vitamin C.

USE OF DRUGS

TABLE 1.1 *(Continued)*

Primary Drug	Drugs that Interact with Primary Drug	Reactions that May Be Produced and Comments
	X-ray to head	Accelerates onset and prolongs duration of action. Physician should know if barbiturates have been taken prior to cephalic X-ray.
	Black Widow spider bite (Black Widow spider venom)	Barbiturates (and morphine) potentiates action of venom and can cause respiratory paralysis. May be lethal. Do not use.
Blood cholesterol lowering drugs. Examples: Clofibrate, Triiodonthyronines, etc.	Anticoagulants	Potentiate one another's effects.
	Antidiabetics (oral) (see Alcohol)	Potentiate one another's effects.
	Oral contraceptives	Antagonize blood cholesterol drugs, reducing their effectiveness.
Caffeine (see CNS stimulant under Alcohol). Examples: coffee, tea, and many analgesics.	Alcohol	See Alcohol under Primary Drug heading.
	Valium (minor tranquilizer)	Reverses effectiveness of this tranquilizer.
	MOA inhibitors (see Alcohol)	Excessive use of caffeine can cause high blood pressure.
	Darvon (narcotic)	Increases effect; fatal convulsions may be produced in overdosage.
Cocaine (see text)	Amphetamines (see text)	Potentiate one another to the degree that vasomotor collapse and respiratory arrest, causing death, is possible.
	MOA inhibitors	Potentiation may be severe enough to cause death.
	Heroin	An additive effect is produced. Unpredictable in reaction of individual.
Cold and cough remedies		See Antihistamines under Primary Drug heading.
Dietary supplements	Contain various combinations of drugs often including:	Study formula and determine whether a potential hazardous drug interaction may occur.

TABLE 1.1 *(Continued)*

Primary Drug	Drugs that Interact with Primary Drug	Reactions that May Be Produced and Comments
	alcohol, calcium, choline, copper, iron, magnesium, manganese, potassium, zinc salts, number of vitamins, etc.	
Digitalis and digitalis glycosides (used in treatment of heart disease). Examples: Acylanid, Davoxin, Digitaline, Gitaligen, etc.	Antacids (see Antacids) and antidiarrheal drugs	See Antacids.
	Anticoagulants (see Alcohol)	Digitalis drugs may counteract effects of anticoagulants. Should be given together with care.
	Cathartics (agents increasing bowel movements). Examples: Cascara, Cellophyl, Dulcolax, Epsom Salt, Milk of Magnesia, etc.	Decrease effectiveness of digitalis drugs by speeding them through intestine. But potentiate absorbed digitalis drugs.
	Insulin	Prolonged insulin usage increases the toxicity of digitalis.
	Magnesium-containing compounds	Potentiate digitalis drugs by increasing sensitivity of the heart muscle; may cause toxicity.
Insulin (injectable antidiabetic drug)	Alcohol	See Alcohol under Primary Drug heading.
	Anabolic steroids (used in producing muscle)	Potentiate the blood-sugar reducing action of insulin.
	Amphetamines	Combination is additive in producing low blood-sugar levels. Also increase metabolism and decreases effectiveness.
	MOA inhibitors	A hazardous combination. May cause shock, coma, and possible death.

USE OF DRUGS

TABLE 1.1 *(Continued)*

Primary Drug	Drugs that Interact with Primary Drug	Reactions that May Be Produced and Comments
	Aspirin	Potentiates insulin, increasing effects.
	Thyroid drugs	May increase the required dosages.
Narcotics (see text). All CNS depressants potentiate one another to dangerous levels of CNS depression.	Alcohol	See CNS depressants, Alcohol.
	Antacids	Decrease urinary excretion and thereby potentiate actions.
	Anticoagulants	Prolonged use of narcotics may increase their actions. Potentially dangerous combination.
	Antidepressants	See Antidepressants under Primary Drug heading.
	MOA inhibitors	Potentiate narcotics to dangerous levels of CNS depression.
Oral contraceptives, pills, or injections containing a combination of estrogens and progesterones.	Androgens (hormones often used in cancer therapy)	Oral contraceptives antagonize anticancer effects of androgens.
	Anticoagulants	Many increase the clotting factors. In some women oral contraceptives act as anticoagulants (decreasing clotting).
	Antidiabetics (see Alcohol)	Oral contraceptives may cause increase in blood-sugar levels, requiring an increase in antidiabetic drugs. Oral contraceptives not recommended for diabetics.
	Antihistamines	May reduce effectiveness of contraceptives by causing enzymatic metabolism.
	Barbiturates (see Barbiturates)	Concern by some physicians of complete ineffectiveness of oral contraceptives in women using barbiturates. Barbiturates increase enzyme activity, which destroys oral contraceptives.
	Smoking	Greatly increases the chance of blood clotting (thromboem-

TABLE 1.1 *(Continued)*

Primary Drug	Drugs that Interact with Primary Drug	Reactions that May Be Produced and Comments
		bolism) in women using oral contraceptives.
	Tuberculin skin test	May be ineffective in women using oral contraceptives.
	Vitamin B$_6$	Women using oral contraceptives may need supplemental B$_6$ if they complain of upper leg aches.
	Black Widow, Scorpian, and other neurotoxic venoms	Narcotics enhance the toxicity of these venoms.
Tyramine(amine) or tyrosine (amino acid)—rich foods. Examples: beer, cheddar cheese (other ripe cheeses), chocolate, fermented products, Chianti wine, meat extracts, yogurt, etc. Tyramine-rich foods instrumental in migraine headaches.	Amphetamines	Combination may cause high blood pressure to severe levels.
	MOA inhibitors	Death has occurred from combinations of these foods with MOA inhibitors. Least reaction may be severe migraine headaches.
	Morphine	Antagonistic pain-killing effects of morphine.
Tranquilizers (minor). (See text for examples). CNS depressants potentiate all other CNS depressants.	Alcohol	Lowered tolerance to alcohol. Severe low blood pressure and sedation may occur.
	Antidepressants (see Alcohol)	Potentiate certain tranquilizers. Many lower CNS to convulsive threshold-producing seizures.
	Antihistamines	Potentiation, both CNS depressants.
	MOA inhibitors	Potentiation may be hazardous.

[a] For further information and interactions, see: Eric W. Martin, *Hazards of Medication* (Philadelphia: Lippincott, 1971).

USE OF DRUGS

According to law, all nonprescription drugs must list these seven points on the label:

1. Name of the product.
2. Name and address of the manufacturer, packer, or distributor.
3. Net contents of the package.
4. Active ingredients and the quantity of certain ingredients.
5. Name of any habit-forming (addicting) drug contained in the product.
6. Cautions and warnings needed for the protection of the user.
7. Adequate directions for safe and effective use.

Remember, labeling and warnings help protect you against misuse and harmful effects; follow them closely (see Figure 1.2).

Proper Limits of Self-Medication

Most often people medicate themselves to relieve symptoms, not to treat a particular disease. This level of knowledge does not often permit a person to recognize the coexistence of these symptoms with a serious underlying disease. People should not self-treat any chronic, persistent, unresponsive, or frequently recurring condition or symptom. Medical advice should be sought to determine the cause of any recurring complaints. A good rule to follow is: Do not treat anything you do not fully understand; instead present it to a physician for medical advice.

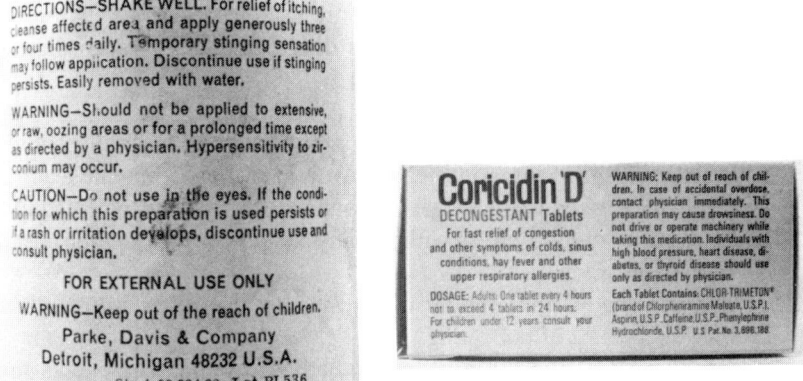

Figure 1.2. Label warnings. The labels on nonprescription drugs should be followed to protect against misuse and harmful effects.

When two drugs are taken together or within a few hours of each other, they may interact, altering the expected action of either drug. These interactions occur between prescription drugs, between over-the-counter drugs, and at times between drugs and foods. Because of the possibility of drug interactions, your physician should always know the names of *all* drugs, both prescription and over-the-counter, that you are using at any one time.

MAJOR ACTIONS AND EFFECTS OF DRUGS

The drugs used in this country are so numerous that it would be almost hopeless to attempt to study them without the aid of some system of grouping. Drugs may be classified in various ways, for example, according to their source, according to their chemical composition, according to their actions and effects, or according to their medical use. None of these methods is entirely satisfactory, but probably the most useful method for the purposes of this book is to classify drugs on the basis of their major actions and effects. Any action or effect of a drug, after it has entered the body, depends upon the dosages used, how and where the effects take place, and the ultimate elimination of the drug by the body.

Drug Dosages

A dosage of a drug is defined as the amount of the drug that is administered at one time. The degree of response depends upon the effectiveness of the drug and the dosage. As will be shown in Chapter 3, the dosages used by an individual are an extremely important part of any study of drug abuse. A number of terms are used to describe the amount of drug in a dosage, and the dosage of a drug is calculated according to the actions it will produce (Figure 1.3).

Minimal dosage is the smallest amount of a drug that will produce a *therapeutic effect* (amount needed to treat or heal someone).

Maximal dosage is the largest amount of a drug that will produce a desired therapeutic effect without any accompanying symptoms of *toxicity* (poisoning).

Toxic dosage is the amount of a drug that produces untoward effects or symptoms of poisoning.

Abusive dosage is the amount needed to produce the effects and actions desired by an individual abusing a drug. This is usually a toxic amount.

Lethal dosage is the amount of a drug that will cause death.

MEDICAL CLASSIFICATION OF DRUG ACTIONS AND EFFECTS

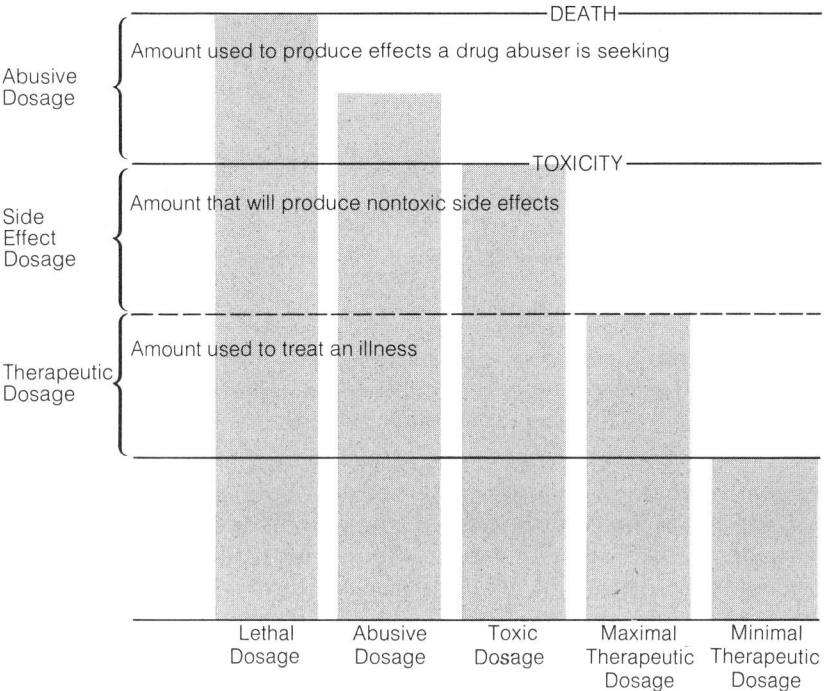

Figure 1.3. Relative dosages of drugs.

MEDICAL CLASSIFICATION OF DRUG ACTIONS AND EFFECTS

Present knowledge of the chemical structure of drugs enables scientists to predict what a drug will do and what the results will be. Drugs act on many parts of the body and in a number of different ways. Certain specific terms are used when referring to those medical actions and effects.

Local Drug Action

Many drugs act only at the site of application, which is usually on the skin or the mucous membranes of the body. This is classified as a *local* action.

Systemic Drug Action

If a drug is absorbed into and distributed through the body, its action is said to be *systemic*. Drugs are most often absorbed and distributed by the blood stream. Systemic action may affect the whole body, or it may be restricted to target organs within the body.

Terms used to describe the major actions and effects of drugs with systemic action are explained in the following paragraphs:

Stimulation. Stimulation is the action of a drug that increases the activity of cells. For example, caffeine is called a stimulant because it increases the activity of cells in the central nervous system. However, prolonged overstimulation of cells can often result in depression.

Depression. Depression is the action of a drug that decreases the power of cells to function. For example, narcotics depress the ability of the respiratory center to send out nerve impulses to the respiratory muscles, thereby reducing respiration.

Selective action. Selective action is the ability of a drug to produce a greater effect on a particular tissue or organ than it does on others. For example, certain posterior pituitary hormones have a highly selective action on the functioning of the tubular cells of the kidney.

Therapeutic action. The action of a drug on diseased tissues or in a sick individual is called therapeutic action.

Side effect. Any effect of a drug other than the one for which it is administered is a side effect. This effect is not necessarily harmful to the individual. For example, morphine is often given to relieve pain, not to constrict the pupil of the eye, which is a side effect.

Untoward effect or action. A side effect regarded as harmful to the individual is an untoward effect or action. For example, the action of morphine that results in nausea, vomiting, constipation, and addiction is undesirable and harmful.

Cumulative action. Cumulative action is the effect that may be produced when drugs are excreted or destroyed more slowly than they are absorbed, so that they accumulate in the body. A sufficiently high concentration may be toxic.

Synergism, potentiation, and additive effects. Synergism, potentiation, and additive effects are ways in which drugs may work together. When drugs that produce the same general kind of effect are given together, they may produce an exaggerated effect out of proportion to that of each drug given separately. Such drugs are said to be *synergistic*. For example, a combination of barbiturates and alcohol produces a total depressant effect much greater than the sum of the two effects when these drugs are taken separately.

When two drugs are given together and one intensifies the action of the other, one drug is said to *potentiate* the other. For example, when epinephrine is given with the local anesthetic procaine, it intensifies the effect of the procaine because it produces constriction of the blood vessels and holds the procaine in the area where its action is desired.

When two or more drugs are given together and their effects represent the action of one plus the action of the other (or others), their effects are said to be *additive*. For example, half a dose of one drug plus half a dose of a similarly acting drug may produce the additive effect of a full dose of either.

Antagonistic drugs. Drugs that produce temporary effects opposite to those usually expected of a given drug are said to be antagonistic. The antagonistic action is valuable in treating toxic overdoses or poisonings. Such is the action of some tranquilizers in relieving the effects of LSD.

Idiosyncrasy. Any abnormal susceptibility of a person to a drug is referred to as an idiosyncrasy. Not only do different individuals have different reactions to a particular drug, but also one individual may react differently at various times to the same drug. This is one of the major reasons why individuals should not prescribe drugs for themselves, use drugs that have been prescribed to treat previous conditions, or use drugs that have been prescribed for others.

Hypersensitivity. An allergic response to a drug is called hypersensitivity. This means that at some time the individual was sensitized to a substance and that thereafter any small dose of the drug can produce a heightened reaction that may vary from slight to severe.

Drug interactions. Drug interactions occur when two drugs are taken at the same time or within a few hours of each other, thus altering the expected action of either drug. Such interactions make up one of the major causes of adverse drug reactions. They may occur between prescription drugs, between prescription and over-the-counter drugs, and at times between drugs and food. For example, in 1963 it was recognized that individuals who were taking monoamine oxidase (MAO) inhibitors could have fatal brain hemorrhages if they ate cheese.

Drug interactions are the reason your physicians should always know the names of *all* drugs, prescription and over-the-counter, that you are using at one time.

Tolerance. Tolerance is an acquired ability to endure, without ill effects, a given dosage of a drug. Such reaction to a drug necessitates an increase of dosage in order to maintain a given effect (see Table 4.1, p. 118). Opiates, barbiturates, and alcohol readily create tolerance. Also, exposure to certain drugs may result in tolerance to chemically related drugs with a similar action. This is known as *cross-tolerance*. Individuals who are chronic heavy users of alcohol have a tolerance to certain anesthetics, such as ether, and require more of the anesthesia than does the nonalcoholic individual.

Habituation. Generally, psychic dependence upon a drug (see Table 4.1) is referred to as habituation. If the psychic influence is strong, individuals may become habituated to almost anything. For example, if a patient believes that a certain little white pill makes it possible to sleep at night, he or she probably will not sleep until the pill has been swallowed. It makes no difference whether the tablet is made of sugar or a strong sedative.

Addiction. Addiction is, generally, physical dependence upon a drug, combined with a psychological dependence (see Table 4.1). It is characterized by a specific series of symptoms that develop because of the altered physiological processes and psychological effects produced by the use of certain drugs.

Table 1.1 outlines some of the major actions, effects, and interactions possible when using more than one substance at one time. Often people do not realize the severity of taking prescription and over-the-counter (nonprescription) drugs together or with certain kinds of foods or alcoholic beverages.

SUMMARY

    I. A drug
       A. Two definitions
          1. Common (public)—a medicine or substance used in the treatment of a disease.
          2. Scientific—any substance, other than food, which when introduced into the body, alters the body or its functions.
       B. The actions and effects of drugs vary widely.

    II. Medical use of drugs
       A. Drugs have been used for thousands of years for healing.
       B. New drugs introduced during the past 30 years have revolutionized the practice of medicine.
       C. Recent "wonder drugs" have helped create unrealistic expectations of drugs and have led to their abuse.

SUMMARY

III. Administration of drugs
   A. Methods (see Figure 1.1)
      1. Oral ingestion (by mouth) and absorbtion from the digestive tract.
      2. Inhaled into the lungs.
      3. Injected under the skin, into a muscle, or a vein.
   B. After being taken into the body, the drug is distributed by the blood stream to the body's organs, tissues, and cells.

IV. Treatment of diseases and conditions
   A. Prescription of a drug can be made only after a diagnosis.
   B. Types of therapy.
      1. Dietary—regulation of intake of fluids and food.
      2. Operative—surgical intervention.
      3. Physical—use of heat, cold, light, radiation, electricity, and massage.
      4. Psychotherapy—counseling to remove or reduce emotional factors in disease.
      5. Chemotherapy—use of drugs to treat disease.
   C. What a patient can do
      1. Follow all instructions on the use of a drug.
      2. Refrain from self-treatment.

V. Names of drugs
   A. The official name is that listed in *The Pharmacopeia of the United States*.
   B. The chemical name provides a structural description of a drug.
   C. The generic name identifies it according to chemical family.
   D. The trademark (brand) name is that name assigned by the manufacturer which may be legally registered and restricted to use by the manufacturer.
   E. Common (street, slang) names are those names assigned by users, who often are engaged in the illegal use of drugs.

VI. Use of drugs
   A. Drugs should be prescribed by physicians only with caution, although some physicians prescribe drugs more readily than do others.
   B. Prescription and nonprescription drugs
      1. Nonprescription drugs are those safe for use without medical supervision, which therefore may be sold over the counter.
      2. Prescription drugs are those not safe for unsupervised use, and which therefore require a label with the legend: "Caution: federal law prohibits dispensing without prescription."

C. Self-medication
  1. Over-the-counter (O-T-C) drugs are also called "patent medicines or "proprietary compounds."
  2. The most common example would be aspirin.
     a. It can produce unwanted effects when taken regularly and in large amounts (see Table 1.1).
     b. Its use with serious conditions may mask symptoms and lead to a more serious condition.
D. Directions for use of nonprescription drugs
  1. Following the warnings found on the label. (See Figure 1.2 for a typical label.)
  2. Use only as directed.
E. Proper limits of self-medication
  1. One should not treat by themselves any chronic, persistent, unresponsive, or frequently recurring condition or symptom.
  2. Be careful when taking two drugs within a few hours of each other.
  3. Keep your physician informed of all drugs, including over-the-counter, you are taking.

VII. Major actions and effects of drugs
  A. Among other methods, drugs may be classified according to their actions and effects (medical use).
  B. Any action depends upon the dosage used.
  C. Drug dosages—the amount of drug administered at one time.
    1. Minimal dosage—the smallest amount of a drug that will produce a therapeutic effect.
    2. Maximal dosage—the largest amount that will produce a desired therapeutic effect without symptoms of toxicity (poisoning).
    3. Toxic dosage—the amount that produces unwanted effects.
    4. Abusive dosage—the amount needed to produce effects and actions desired by a person abusing the drug.
    5. Lethal dosage—the amount that will cause death.

VIII. Medical classification of drug actions and effects
  A. Drugs act on many parts of the body and in a number of ways.
  B. Local drug action—any drug that acts only at the site of application.
  C. Systemic drug action—any drug absorbed into and distributed through the body, usually by the blood stream. Action may be:
    1. Stimulation—an increase in the activity of cells.
    2. Depression—a decrease in the activity of cells.
    3. Selective action—the ability of a drug to produce a greater

SUMMARY

effect on a particular tissue or organ than it would on other tissues or organs.
4. Therapeutic action—the effect on diseased tissues.
5. Side effects—any effect other than the one for which it is administered.
6. Untoward action—a harmful side effect of a drug.
7. Cumulative action—the effect produced when drugs are excreted or destroyed more slowly than they are absorbed, so that they accumulate in the body.
8. Synergism, potentiation, and additive effects—when drugs act together to produce an exaggerated effect out of proportion to the effect of each drug individually.
9. Antagonistic drugs—drugs that produce temporary effects opposite to those usually expected from that drug.
10. Idiosyncrasy—an abnormal susceptibility of a person to a drug.
11. Hypersensitivity—an allergic response to a drug.
12. Drug interactions—effects produced when two drugs are taken at the same time or very close together, thus altering the expected action of either drug.
13. Tolerance—ability to endure, without ill effects, a given dosage of a drug.
14. Habituation—psychic dependence upon a drug.
15. Addiction—both a physical and psychic dependence upon a drug.

NEUROLOGICAL ASPECTS OF ABUSIBLE DRUGS

Humans consist of cells organized into tissues; the tissues into organs; and the organs into systems (Figure 2.1). All of these working as a unit provide humans with their vital functions. These functions need coordination and direction to work as a whole. The main coordinator of body functions is the *nervous system,* which consists of the *central nervous system* (brain and spinal cord) and the *peripheral nervous system* (all of the nerves extending from either the brain or spinal cord). The basic functional unit of the whole nervous system is the *neuron* or nerve cell.

When neurons transmit a nerve impulse to one another, to a muscle, or the muscular tissue in the walls of an organ, the first neuron does not actually touch the second neuron or the muscle cell. The junction between two neurons is a *synapse,* and the junction between the neuron and the muscle tissue is termed the *neuromuscular junction.* Transmission of a nerve impulse across these junctions is done by production of a specific chemical (a *neurotransmitter*) at the end of the first neuron which stimulates the second neuron or the muscle cell and is *deactivated* at the same time by a specific enzyme produced at the point where the second neuron or muscle cell was activated. This allows nerve impulses to cross a synapse or neuromuscular junction *one at a time.* Also, such a chemical transmission system allows nerve impulses to be transmitted in only one direction.

The major neurotransmitter chemical in the nervous system is *acetylcholine* (ACh). Nerve fibers that produce acetylcholine as their neurotransmitter are termed *cholinergic fibers.* Another group of neuro-

NEUROLOGICAL ASPECTS OF ABUSIBLE DRUGS

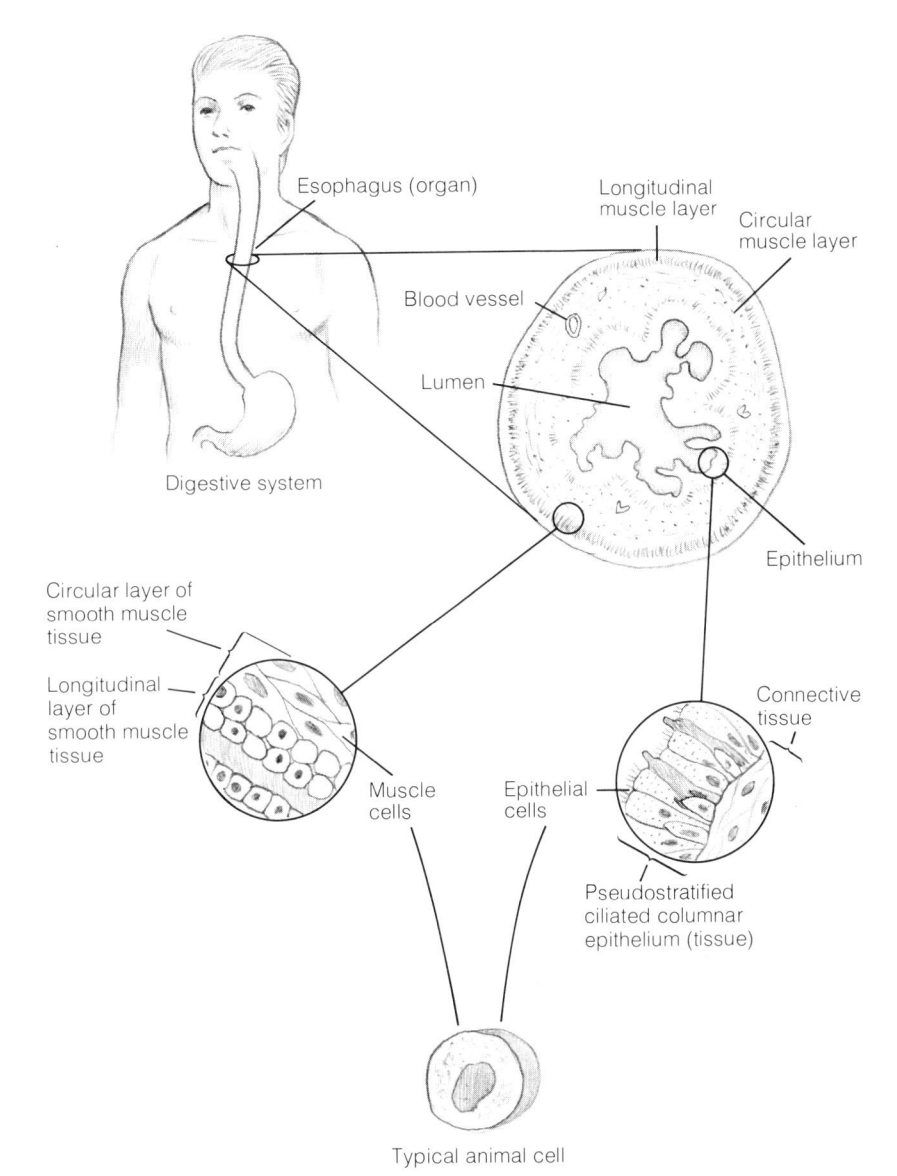

Figure 2.1. Cell–tissue–organ system. The digestive system is used as an example of the level of organization in animals such as human beings.

transmitter chemicals are termed *catecholamines*. These include: *dopamine*, *norepinephrine* (noradrenalin), *epinephrine* (adrenalin), and a synthetic chemical *isoproterenol* (isoprenaline). Fibers producing a catecholamine as the neurotransmitter chemical are called *adrenergic* fibers. A third neurotransmitter, functioning mainly in the brain, is the monoamine neurohormone, *serotonin.*

The central nervous system functions much like a computer, and the peripheral nervous system like a telephone transmission line. Information from the external environment (sights, sounds, tastes, etc.) and from inside the body (blood pressure, body temperature, heart beat, muscle tension, etc.) is relayed by the peripheral nervous system to the appropriate part of the central nervous system. This information is recorded, stored (memory), coordinated with other information, and instructions are relayed to the muscle cells of the organs and systems to produce the necessary actions and adjustments to the body. In turn, these adjustments are fed back into the central nervous system. Such a constant flow of information permits continuous adjustment and control of all human functions, both physiological and psychological.

Research into drugs with a potential for abuse is concerned with their biochemical actions within the nervous system. Researchers attempt to establish the chemical characteristics of certain drugs and their links with the mood, personality, and behavioral modifications that individuals exhibit when using such drugs. These drugs influence specific neurons in the nervous system and either increase or decrease the activity of neurotransmitters, conducting nerve cells, or specific brain centers. The physical effects of such drugs are very predictable, but the mood and behavioral responses depend upon the individual's personality, emotional state, and a number of other factors. Because of the mood modifications and behavioral changes produced in some individuals, such drugs are called *mood-modifiers, psychotropic, psychoactive,* or *psychotoxic* substances.

HOW PSYCHOACTIVE DRUGS WORK

A drug may act on the surface of cells or within cells (as enzymes or on other specialized cellular components). Drugs are selective and exert their actions by becoming attached to specialized chemical groups or parts of a cell membrane called *receptor sites.* The relationship of a drug to its *receptor* is pictured as being like the fit of a key into a lock. The receptor-binding of the drug to a specific cell then enables the drug to participate in a few steps of the normal sequence of a cellular process. In this manner substances may interfere with, alter, or replace chemicals of normal cellular processes. Some of these actions seem relatively simple and easy to understand; some are exceedingly complex; and many are at present unknown.

Some substances bind to a receptor, keeping a needed chemical from entering a cell and blocking a cellular manufacturing process, a storage process, a releasing process, or a transmitter system. Other drugs exert their effects by binding to chemicals within a cell membrane and altering the cell's function, either by reducing the membrane's permeability or increasing it's permeability to specific substances. Drugs binding to cell-membrane lipid (fats) receptor sites can cause decreased oxygen entrance into the cell, resulting in depressed cellular activity. Receptors that obstruct or alter normal biochemical sequences within a cell may also lead to a piling up of usually insignificant products, which often produces stimulating effects.

Many psychoactive drugs exert their effects upon the synapses and the neurotransmitter chemical of the synapses. The differences between the drugs are in where they act (what part of the brain, spinal cord, etc.) and how they act (either stimulating or inhibiting the synaptic events).

A drug that combines with the same receptor site as another drug but does not initiate the same drug actions is an *antagonist*. Antagonists work by tying up specific receptor sites, thus preventing access and stopping the action of the original drug. This antagonistic action is valuable in treating toxic overdoses or poisonings. Such is the action of some tranquilizers in relieving the effects of hallucinogens and other drugs.

Drug Elimination Within the Body

A drug introduced into living cells is subject to the same biochemical transformation and elimination from the cells as a normal cell product. Drugs may be chemically inactivated, excreted from the body unchanged, or chemically transformed into less effective substances or waste products. The liver is the organ of the body that most often transforms drugs. Often, the effects of a drug upon the body are directly related to the liver's ability to metabolize that drug. If such a drug is to have an effect, this ability of the liver must be overcome. The liver mechanism helps the body to tolerate the effects of some drugs without producing toxic effects to the individual or damage to the body. With other drugs, this ability to tolerate takes place within the cells being acted upon. The ability of the body to overcome or reduce the effects of certain drugs is termed *tolerance*. With some drugs (amphetamines, alcohol, narcotics, etc.), the potential for body damage is so great that supplemental systems for drug metabolization develop in the liver, greatly reducing the effectiveness of the drugs.

Figure 2.2 describes an experimental demonstration of the development of drug tolerance. A group of individuals were given equal dosages of heroin every day for 19 days (*Equivalent Daily Dosage* line). The *euphoric* effects (exaggerated sense of well-being) they experienced after receiving the heroin were measured each day and graphed as the *Tolerance* line. As you can see, the effects of the standard dosage decreased and by the nineteenth day were almost nonexistent, even though

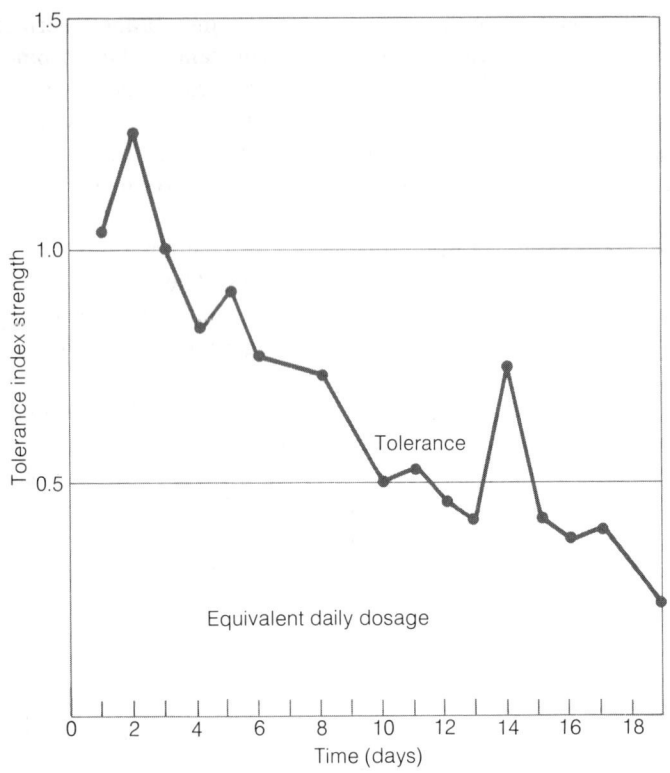

Figure 2.2. An experimental demonstration of drug-tolerance development conducted at the National Institute of Mental Health Addiction Research Center in Lexington, Kentucky. Heroin users respond to the development of tolerance by progressively increasing their dosages of heroin. (Adapted from W. R. Martin and H. F. Fraser, "A Comparative Study of Physiological and Subjective Effects of Heroin and Morphine Administered Intravenously in Postaddicts," Journal of Pharmacology and Experimental Therapeutics, *1961, p. 397.) Copyright © 1961 by Williams and Wilkins Co., Baltimore, Maryland.*

the same chemically measurable amount of heroin was given on the nineteenth day as on the first.

This graph demonstrates that tolerance to heroin develops very quickly, greatly reducing the euphoric effects of the heroin upon the drug abuser. If tolerance had not developed, the Tolerance line would have remained equal to the Equivalent Daily Dosage line throughout the 19 days. Heroin users respond to the development of tolerance by progressively increasing the dosages they take. This way they are able to overcome tolerance and experience the same euphoric effects, or "high," every day.

Whenever tolerance develops, an individual must progressively increase the dosage he or she is taking in order to produce the desired effects. Such action may result in cellular damage, quite often in the liver (an organ necessary for life).

Most drugs disappear from the blood and site of action at a rate proportional to the concentration at any given moment. When concentration is high, the rate of disappearance is high; when concentration is low, the rate of disappearance is low. Destruction or removal of drugs from the body begins immediately upon their introduction into the body. The rate of drug removal by excretion, diffusion into tissues, and chemical alteration within the body is also dependent upon the blood concentration. Drugs that are not stored cannot be maintained in the blood or tissues. When the amount of a drug in the blood falls below the excretion abilities of the body, the excretion action ceases.

Elimination of drugs from the body in either an active or an inactive form occurs by way of the usual channels of excretion. The kidney is the most important organ of excretion. Urine, bile, sweat, saliva, milk, tears, and expired air may contain drugs or the degraded (chemically simplified) products of drugs. Some drugs are also excreted directly into the large intestine.

DRUG-INDUCED EFFECTS ON THE CENTRAL NERVOUS SYSTEM

Mood-modifiers, psychotropic, psychoactive, or psychotoxic substances bind to specific receptor cells in the central nervous system, producing their effects. Such drugs affecting the brain alter the consciousness of an individual in a number of ways. The most significant include: disturbances in thinking (altered attention, memory, or judgment); changes in perception of time or space (time may seem speeded up or slowed down or become suspended and space may increase, decrease, or cease to exist); loss of control (feelings of helplessness, inability to change body positions); alterations of emotions (euphoria, outbursts of tears or laughter); perceptional distortions (illusions or hallucinations —perceptions of objects with strange and unusual relationships to reality); or other recognizable changes in personality or behavior. The problem is that it is not known whether the drugs "cause" these effects or "allow" them to take place because of interactions between the physical effects on the brain, the individual's personality, and environment.

The overall actions and effects of these drugs are those of either a stimulant or depressant acting directly upon the central nervous system. A central nervous system *stimulant* is defined as a drug that temporarily increases body function or nerve activity. At times, stimulant drugs produce dramatic effects, but their medical usefulness is limited because of the complexity of their reactions and the nature of their *untoward effects*. Also, repeated administration or large doses may produce convulsive seizures, alternating with periods of depression that may range from exhaustion to coma.

Depressants have the ability to decrease a body function or nerve activity temporarily. Drug-induced depression of the central nervous system is frequently characterized by lack of interest in surroundings, inability to focus attention on a subject, and lack of motivation to move or talk. The pulse and respiration become slower than usual, and as the depression deepens, sensory perceptions diminish progressively. Psychic and motor activities decrease; reflexes become sluggish and finally disappear. If a stronger depressant is used, or if larger (abusive) doses are consumed, depression progresses to drowsiness, stupor, unconsciousness, sleep, coma, respiratory failure, and death.

The nervous system consists of several distinct functional components (see Figure 2.3) fulfilling different psychological functions. Consequently, it is quite natural to find that the drugs which act on the system do not act on all parts of it with the same degree of intensity. Remember, the specific response of an individual to a drug depends to a large extent on the personality of that individual and a number of other social factors as well as on the nature of the drug itself.

STRUCTURAL COMPONENTS OF THE CENTRAL NERVOUS SYSTEM

Various structures of the nervous system regulate specific body functions or store specific information vital to the normal functioning of the body. These structures, when stimulated, produce specific reactions within the body.

Cerebrum

The largest and most prominent part of the brain is the cerebrum, which is divided into the two cerebral hemispheres. The outer surface layer of the cerebral hemispheres is known as the *cerebral cortex* (see Figures 2.3 and 2.4). The cerebral cortex is a vast information-storage area. It is the site of conscious sensations and mental abilities such as memory, intelligence, imagination, creative thought, and recognition.

The sites of conscious sensations are divided into sensory, motor, and association areas. The sensory areas receive information such as sight, smell, taste, hearing, touch, temperature (heat and cold), and pain. The motor areas relay the individual's responses to these sensations to the appropriate muscle or part of the body initiating movement and body changes. The association areas convey both sensations and movements to the mental abilities areas for recognition and human interpretation. As shown in Figure 2.3, the sensory areas are stimulated by specific nerve tracts conveying nerve impulses from lower areas

STRUCTURAL COMPONENTS OF THE CENTRAL NERVOUS SYSTEM

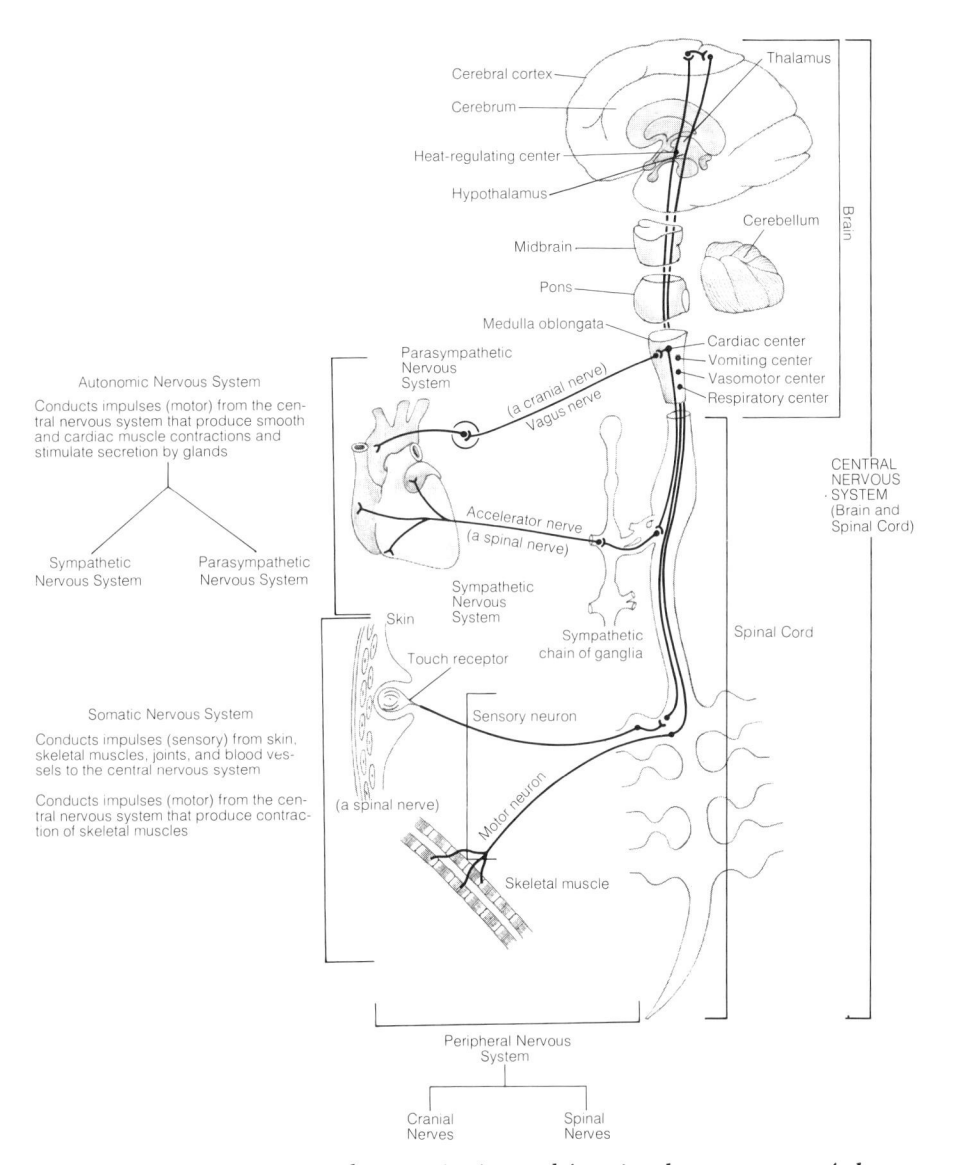

Figure 2.3. Diagram of the organization and functional components of the nervous system.

(brain stem and spinal cord) while other specific tracts transmit impulses back down the nervous system. Directly beneath the cerebral cortex are the nerve tracts of the cerebrum connecting the lower centers of the brain, spinal cord, and all associated areas of the cortex.

When sensory areas of the cortex are stimulated, more numerous

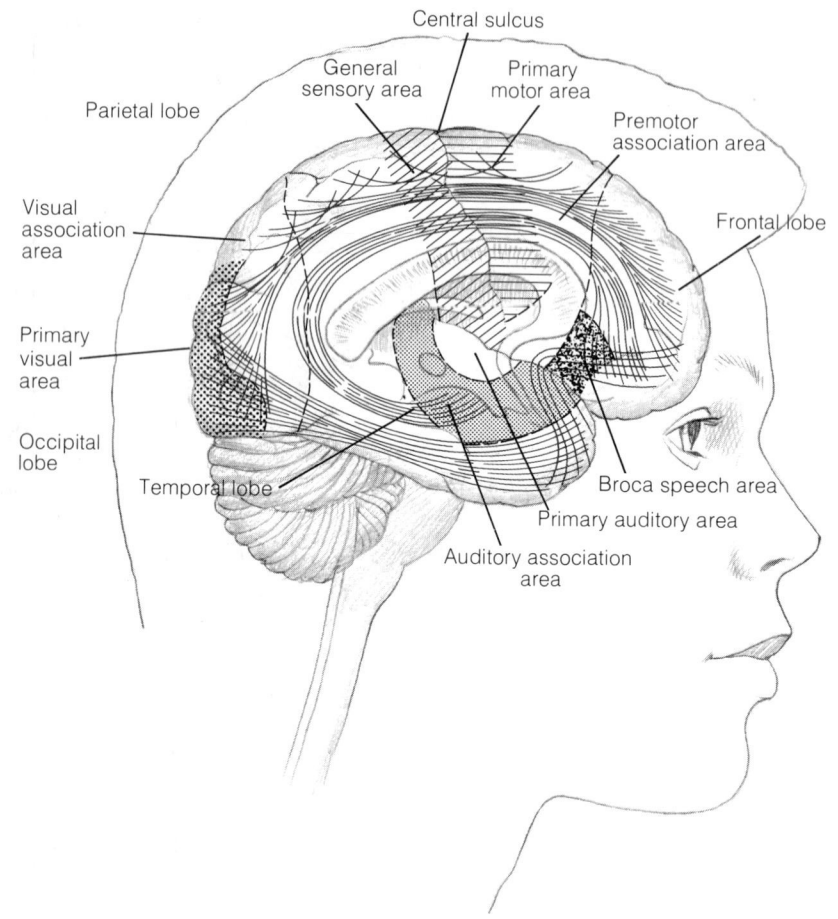

Figure 2.4. Sensory, motor, association tracts, and other areas of the cerebral cortex. (Broca speech area is in the left hemisphere of most people.)

and more vivid impulses are received, and the individual is more alert, more responsive, and more aware of his or her surroundings. When the motor centers are stimulated, a person is likely to be more active and restless; but when overstimulated (which occurs with the abuse of drugs), coordination may be lost, and convulsions may result. Cortical stimulation may make someone more talkative, but if the stimulation is excessive, speech becomes incessant and incoherent, and the individual becomes delirious. Drugs stimulating specific areas of the cerebral cortex (thinking, reasoning, judgment, will, imagination, attention, etc.) may enable the individual to think faster (not necessarily clearer) and form quicker judgments. But at the same time, overstimulated sensory areas may produce illusions and flights of imagination. These sensations and motor actions are also evident when lower areas such as the recticular formation, or limbic system, are stimulated and the stimulation relayed to the cerebral cortex.

Thalamus

The thalamus (see Figures 2.3 and 2.5) serves as a center for impulses to and from the cerebral cortex. It can relay diffuse signals from the brain stem to all parts of the cerebral cortex or cause generalized activation of the cerebrum. The thalamus also can cause activation of specific areas within the cerebral cortex, distinguishing them from other areas. These activation sections of the thalamus are part of the reticular activating system, or limbic system (Figure 2.5), which will be discussed in the next sections. The generalized stimulation of the cortex plays an

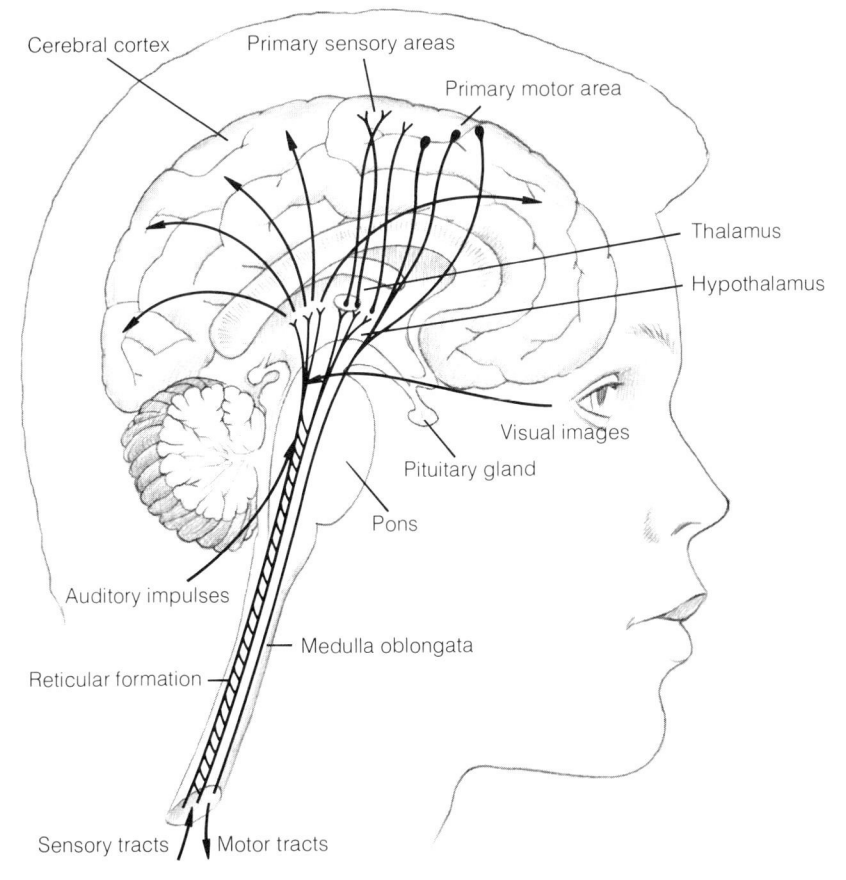

Figure 2.5. Reticular activating system (reticular formation) showing sensory tracts, impulses, and images which activate the system and motor tracts which activate the body.

important role in wakefulness and attention. The selective activation of specific areas probably plays an important role in an individual's ability to direct attention to certain tasks or specific mental activities.

The thalamus is also composed of sensory areas that serve as a center for pain. It also gives the individual impressions of the agreeableness or disagreeableness of a sensation.

Drugs that depress cells in the various portions of the thalamus interrupt the free flow of impulses to the cerebral cortex. In this way pain can be relieved by drugs. Stimulation of the thalamus actively increases the action of the cerebral cortex. Below the thalamus lies the hypothalamus. These two regions are important regulatory centers of the nervous system.

Hypothalamus

The hypothalamus is one of the most important areas of the brain. It controls the basic life functions (regulation of blood pressure, body-fluid balance, appetite, sleep–wake mechanisms, body temperature, and many hormonal secretions of the endocrine glands) and is the mechanism for integrating all these basic functions with the emotions and behavior of an individual. This is done through its relationships with other parts of the nervous system, such as the reticular formation (Figure 2.5), and with the hormones of the endocrine system. Consequently, the autonomic functions and the behavior functions under the control of the hypothalamus are tightly woven into the overall reticular activating system. This interwoven condition gives rise to the body changes associated with emotional changes.

Cerebellum

The cerebellum is the center for muscle coordination, equilibrium, and muscle tone. It receives impulses from the reticular formation and organs of equilibrium (semicircular canals), as well as from the cerebrum. The cerebellum also plays an important role in the maintenance of posture. Drugs that disturb the cerebellum usually cause dizziness and loss of equilibrium.

Medulla Oblongata

The medulla oblongata contains the so-called "vital" centers: the respiratory, vasomotor, and cardiac centers (Figure 2.3). If the respiratory center, for example, is stimulated, it will discharge an increased number of nerve impulses over nerve pathways to the muscles of respiration, thereby increasing respiration. If this center is depressed, it will discharge fewer impulses, and respiration will be correspondingly reduced. Other centers in the medulla that respond to certain drugs are the cough center and the vomiting center. The medulla, pons, thalamus, reticular formation, hypothalamus, and other areas of the midbrain constitute the brain stem and contain many important correlation centers (other than those already described) as well as ascending and descending nerve pathways.

Spinal Cord

The spinal cord is the center for the reflex activity and the transmission of impulses to and from the higher centers in the brain. It may be affected indirectly by the drug actions on higher centers or directly by drugs such as spinal stimulants *(spinal action)*, which may cause convulsions when given in large doses or may just increase excitability when given in smaller doses. [When a drug is described as having *central action*, this means that it has an action on the brain and the spinal cord.]

FUNCTIONAL COMPONENTS OF THE CENTRAL NERVOUS SYSTEM

These are functioning systems or "circuits" which transmit through a number of specific structures of the brain. They are mediating components that regulate the degree of response of the body and brain to stimuli being transmitted within the nervous system. Components such as the *reticular formation, limbic system, Papez circuit,* and *autonomic nervous system* integrate nervous impulses with behavioral responses. Any nerve impulse can be modified by an individual's heredity, learned behavior, and the specific circumstances being experienced at the moment.

Reticular Formation

The reticular formation extends from the spinal cord through to the cerebral cortex. As shown in Figure 2.5, it passes through the medulla, pons, midbrain, and thalamus into the cortex. The senses of the body are carried by specific nerve tracts from lower areas while other nerve tracts carry the impulses back down to the muscles (motor tracts). All sensory and motor pathways carrying impulses to and from the cerebral cortex give off branches into the reticular formation. Consequently, the reticular formation is stimulated whenever information is being transmitted to and from the cortex.

Arousal reactions (wakefulness and alertness). The arousal response results from the sensory activation of the *reticular activating system,* which must be stimulated into action by signals from other areas. When an individual is asleep, the reticular activating system is almost dormant (whereas the cerebral cortex is actively sorting out information accumulated throughout the day). Yet any type of sensory signal will immediately activate the system. For instance, proprioceptive (sensations concerning movements and positions of the body) signals from the muscles, pain impulses from the skin, visual signals from the eyes, auditory signals from the ears, or even sensations from the intestinal organs can cause sudden activation of the reticular activating system of the reticular formation and arouse the individual. This is called the *arousal reaction,* and the reticular formation is called the *arousal center.*

Anatomically, the reticular formation of the brain stem is well constructed to perform such arousal functions. It receives tremendous numbers of collateral fibers from a number of sensory areas of the body. Almost any sensory stimulus can activate it. In addition, many fibers pass directly from the spinal cord to the reticular formation. It, in turn, can transmit signals both upward into the brain and downward into the spinal cord. Many of the fibers originating from cells in the reticular formation divide, with one branch of the fiber passing directly upward and another branch passing directly downward.

The cerebral cortex can also stimulate this arousal system and increase its degree of activity. Direct fiber pathways pass into the reticular activating system from almost all parts of the cerebral cortex. Intense activity of any part of the cerebrum activates the reticular activating system and is usually associated with a high degree of wakefulness. For example, a large number of nerve fibers pass from the motor regions of the cerebral cortex to the reticular formation, and motor activity, or body movement, is closely associated with wakefulness. This helps to explain the importance of movement in keeping a person awake. Stimulation of specific areas of the hypothalamus greatly excites the reticular activating system, causing wakefulness, alertness, and excitement. Such increased stimulation also excites the *sympathetic nervous system* (Figure 2.6), increasing the arterial blood pressure and causing pupillary dilation and general excitement throughout the organs controlled by sympathetic activity. Also, stimulation of other areas of the hypothalamus or isolated areas of the thalamic portions of the reticular activating system (limbic system) often inhibits this excited state, causing drowsiness and sometimes actual sleep through the *parasympathetic nervous system.*

Thus, the reticular activating system (directly) and the hypothalamus and thalamus (indirectly) contribute greatly to the control of the degree of excitement and alertness a person feels.

Motor control (control of muscles). The reticular formation does not act as one large unit; it has the capability of controlling many discrete functions. Diffuse stimulation within it can cause a general in-

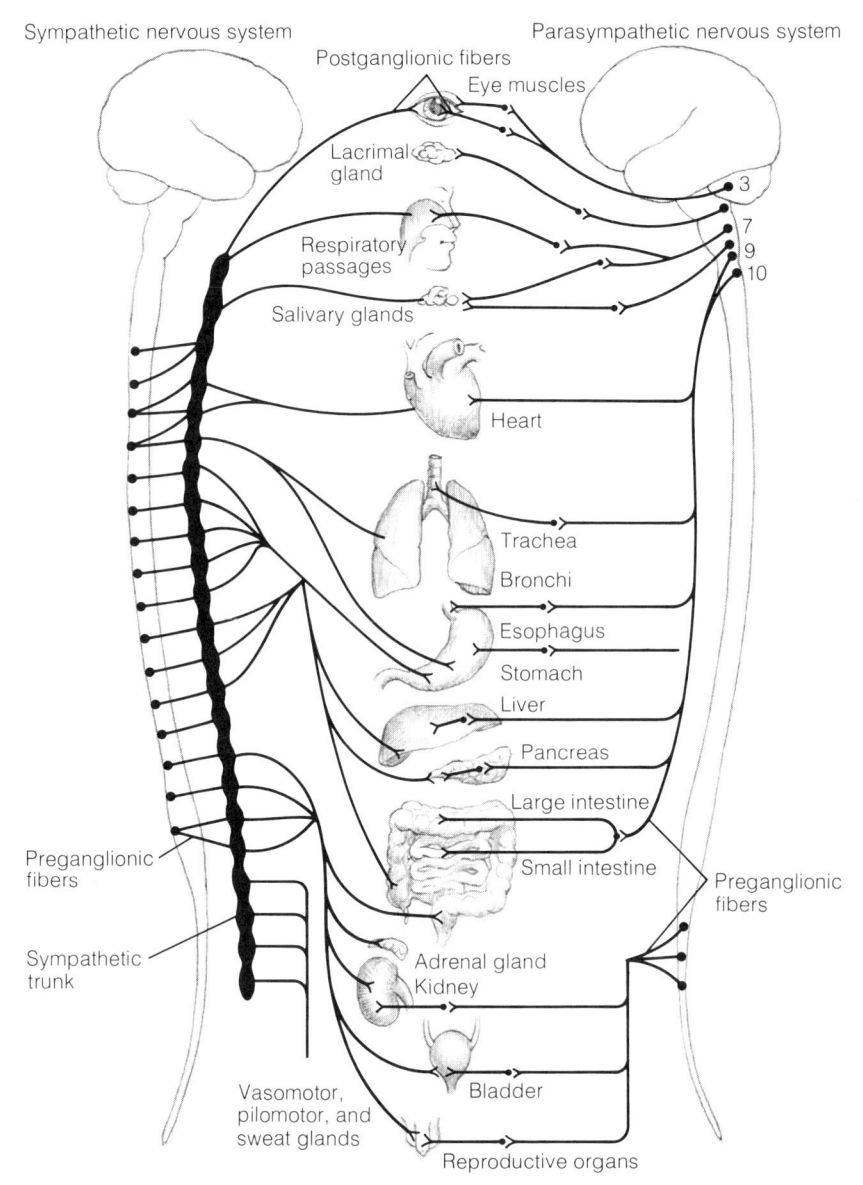

FUNCTIONAL COMPONENTS OF THE CENTRAL NERVOUS SYSTEM
41

Figure 2.6. Diagram of autonomic nervous system showing preganglionic and postganglionic fibers.

crease in muscle tone either throughout the body or in localized areas. Stimulation of discrete points in any portion of the reticular formation will at times cause discrete muscle contraction or inhibition of a muscle contraction in specific parts of the body. Specific areas control the excitation of an *agonist muscle* (a prime, paired muscle with specific functions) along with simultaneous inhibition of an *antagonist muscle* (a muscle with action opposing its paired agonist muscle). This function is extremely important in any body movement or the control of equilibrium and the support of the body against gravity.

General depression of the reticular formation produces sedation and loss of consciousness. Overstimulation distorts sensations and increases motor activity control systems of the body. Many drugs exert effects upon the reticular system. It is particularly sensitive to certain depressants, such as barbiturates and alcohol.

Limbic System

Surrounding the thalamus, along the sides of the lateral fissure, is a group of structures that are collectively referred to as the *limbic system:* hippocampus, fornix, amygdaloid nucleus, hippocampal gyrus, cingulate gyrus, and incus (Figure 2.7). The limbic system and its interconnections with the hypothalamus and thalamus are known as the *Papez circuit* Figure 2.7). The Papez circuit functions as a balancing system for the conscious experience of emotions and emotional expressions. Drugs that interrupt or cause this circuit to become unbalanced can change an individual's perception, expression, and relative level of emotions.

Integrative Reactions

Any sensation or stimulus coming into the brain causes the hypothalamus to send activating impulses back to the body. These impulses controlling the automatic and involuntary functions of the body are transmitted through a special division of the nervous system called the *autonomic nervous system*. This system is divided into two major subdivisions: the *sympathetic* and the *parasympathetic* divisions (Figure 2.6). The nerves of these two divisions are antagonistic to one another. Stimulation of the parasympathetic division is a stabilizing force that generally reduces activity or stabilizes body actions. Stimulation of the sympathetic division increases energy output, increasing organ actions needed for increased effort. This is why the sympathetic division is the emergency or the fight-or-flight division.

These subdivisions of the nervous system conduct nerve impulses from the central nervous system directly to the appropriate organ (Figure 2.6). Also, the reactions of the organs are reported back by sensory nerves and activate the brain, causing conscious and unconscious changes in the body and behavior.

As shown in Figure 2.6, the nerves that transmit messages from the central nervous system through the autonomic nervous system to the organs of the body always consist of two neurons connected in series.

FUNCTIONAL COMPONENTS OF THE CENTRAL NERVOUS SYSTEM

The neuron that conducts the nerve impulse from the central nervous system to synapse with the initial *nerve cell-body* or *ganglion* (a group of nerve cell-bodies outside of the central nervous system) is called the *preganglionic* fiber. The neuron transmitting the nerve impulse on to synapse with the organ effected is called the *postganglionic* fiber.

All preganglionic fibers of the autonomic nervous system (both sympathetic and parasympathetic) release *acetylcholine* (ACh) as the neurotransmitter at the ganglion where the first synapse occurs. Postganglionic fibers of the parasympathetic division also release acetylcholine as their neurotransmitter. These fibers are called *cholinergic fibers*. Postganglionic fibers of the sympathetic division release both acetylcholine and norepinephrine (noradrenalin). Many researchers feel that the release of norepinephrine is caused by the presence of the acetylcholine and that norepinephrine is actually the neurotransmitter

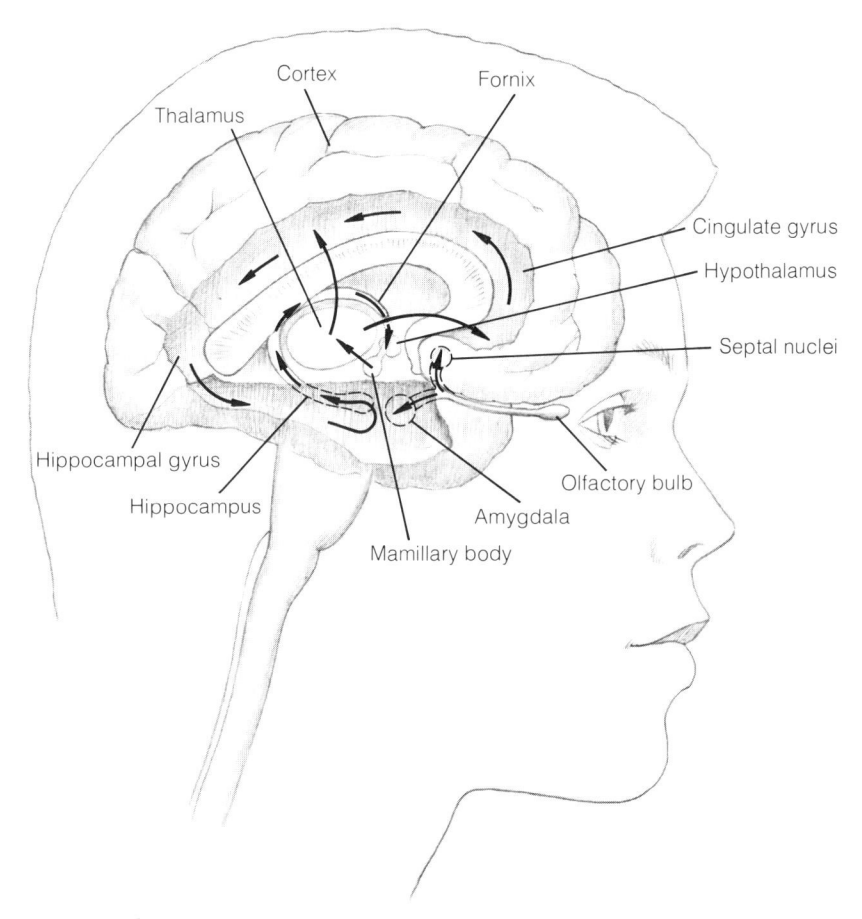

Figure 2.7. Papez circuit.

chemical at these synapse. Fibers liberating norepinephrine as the neurotransmitter are called *adrenergic* fibers.

Drugs that produce effects on the body similar to the parasympathetic division of the autonomic nervous system (stabilizing, constructive, and reparative) are actually acting in place of acetylcholine as transmitter substances at the neuromuscular junction of the preganglionic neuron. These drugs are called *cholinergic drugs*. In the past they were referred to as *parasympathomimetic drugs*.

Drugs that produce effects similar to the sympathetic division (emergency, fight-or-flight response) are replacing the norepinephrine as the transmitter substance at the neuromuscular junction of these postganglionic neurons. These drugs are called *adrenergic drugs;* the older term for these was *sympathomimetic drugs*.

In recent years there has been increasing evidence to connect the hypothalamus, reticular activating system, limbic system, and their transmission of nervous impulses through the *sympathetic* and *parasympathetic divisions of the autonomic nervous system* with the regulation and coordination of levels of behavior, moods, emotions, and changes in body functions associated with these states. Also, the abilities of certain psychoactive drugs to change and regulate moods, emotions, and behavior have been linked to the functioning of these areas and systems.

Scientific evidence supporting the fact that behavioral states (and side effects) are controlled by two separate and antagonistic systems (divisions) functioning within the hypothalamus is accumulating. As shown in Figure 2.8, they are the *ergotropic division,* which functions through the sympathetic nervous system, and the *trophotropic system,* which functions through the parasympathetic nervous system.

All functions of the central nervous system are dependent upon the actions of certain chemicals called *monoamines, neurohormones, catecholamines,* or *neurotransmitter substances,* which are produced by the body or brain and stored in the brain in inactive forms. At the right moment they are released in an active form in correct amounts to allow transmission of a nerve impulse across the synapse in a specific nerve pathway or to inhibit the synaptic transfer of the nerve impulse. These actions either produce a stimulant or depressant effect or allow a stimulant or depressant effect to take place. *Acetylcholine* has been identified as the neurotransmitter in the central control areas of the hypothalamus and related areas (Figure 2.8). *Norepinephrine* is the neurotransmitter of the ergotropic division, and *serotonin* is the neurotransmitter for the trophotropic system (Figure 2.8).

The ergotropic division is thought to integrate the sympathetic nervous system and reticular activating system with the senses (stimuli) and the voluntary muscular activity of the body. Norepinephrine is the neurotransmitting chemical, and release of it produces increased emo-

FUNCTIONAL COMPONENTS OF THE CENTRAL NERVOUS SYSTEM

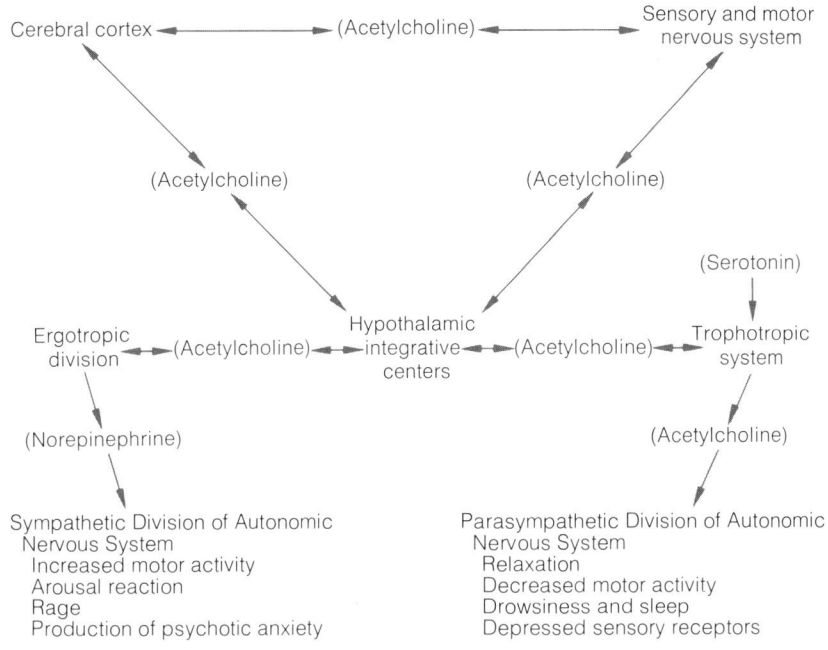

Figure 2.8. Hypothalamic integrative mechanisms. These indicate a hypothetical balance between the ergotropic and trophotropic systems. The degree of behavioral responses is dependent upon the amount and/or frequency of stimulation of this system. [Adapted from Betty S. Bergersen, and Andres Goth, Pharmacology in Nursing, *12th ed. (Saint Louis: C. V. Mosby, 1973), p. 34, fig. 16-3]*

tional responses, increased reaction to sensory stimuli, arousal and excitement, increase in muscular control, and an increase in any functions under the control of the sympathetic division of the autonomic nervous system. Thus, stimulation of this system in the hypothalamus produces an activated psychic and physical state. When regulated, these combined responses seem to ready the body for positive action. When overstimulated, they can produce behavior seen in psychotic anxiety states and certain drug reactions.

The trophotropic system seems to integrate the parasympathetic division, the limbic system, and the senses, and the voluntary muscular activity. Serotonin neurotransmission activates this system and decreases reaction to stimuli, causing decreased psychomotor activity; overstimulation causes drowsiness or sleep. This is responsible for such behavioral patterns as relaxation, subdued psychic states, and lack of attention found in certain emotional responses and drug reactions.

The degree of alteration in psychomotor activity, behavior patterns, and emotional responses depends upon the balance of serotonin and norepinephrine in the hypothalamus. Drugs that mimic their actions or that cause the overproduction of them can let these chemicals build up and cause changes in behavior. There are also systems of *enzymes* (chemicals that facilitate chemical changes in the body) that break down or deactivate neurohormones. *Monoamine oxidase* (MAO) is an enzyme that either destroys serotonin and norepinephrine or causes them to be stored in inactive forms. Blockage of this enzyme can cause them to build up, producing changes in behavior.

Pain and pleasure (reward and punishment). The hypothalamus and closely related structures such as the reticular formation are greatly concerned with the effectiveness of sensory sensations, that is, whether the sensations are pleasant or painful. These qualities are also called *reward* and *punishment*.

The major centers controlling pain and pleasure have been located in the hypothalamus. These areas are also the most reactive of all the hypothalamic areas. It also seems that stimulation in the pain centers can frequently inhibit the pleasure centers completely. This explains why pain can often take precedence over pleasure.

Rage (affective-defensive pattern). Stimulation of the regions of the hypothalamus that give the most intense sensations of punishment also causes a peaceful animal to develop a classic emotional pattern called *rage*—the assumption of a defense posture such as extended claws, lifted tail, hissing, spitting, growling, or wide-open eyes with dilated pupils. Humans, however, generally indicate rage with extended arms, throat sounds, and wide-open eyes with dilated pupils. The slightest provocation, while the individual is in this state, causes an immediate, savage attack.

Exactly the opposite emotional behavior pattern occurs when the pleasure centers are stimulated, namely, relative docility and tameness.

Psychological and psychosomatic effects. Abnormal psychic states can greatly alter the degree of nervous stimulation to the skeletal muscles throughout the body. This can increase or decrease the skeletal muscle tone. In neurotic and psychotic states such as anxiety, tension, and mania, generalized overactivity of both the muscles and the sympathetic nervous system often occurs throughout the body. The hypothalamus and reticular activating system undoubtedly help to maintain the extreme degree of wakefulness and alertness that characterizes those emotional states. Also, the docility and somnolence characteristic of extreme depression show the abilities of the hypothalamus to decrease the muscle tone throughout the body.

Many psychosomatic abnormalities (such as ulcers and rashes) result from hyperactivity of either the sympathetic or the parasympathetic system. Emotional patterns controlling the sympathetic and parasympathetic centers of the hypothalamus can cause wide varieties of peripheral psychosomatic effects.

SPECTRUM AND CONTINUUM OF DRUG ACTIONS

The degrees of depression and stimulation produced by drugs affecting the central nervous system are not discrete actions. As shown in Figure 2.9, the spectrum of drug actions can be set into a continuum of effects and actions.

The continuum of drug effects extends from overstimulation of the nervous system to death at one extreme and from severe depression of the nervous system to death at the other extreme. The central or neutral area of this continuum is the degree of stimulation and depression usually encountered in everyday living.

The drug groups are placed along the continuum according to the actions or effect that they produce when normal therapeutic (minimal) dosages are consumed by an individual. Drugs in different groups may have similar side (or untoward) actions even though their major actions differ. For example, narcotics are used to relieve pain (major actions), but they may also cause a person to be drowsy or sleepy (side action). Barbiturates are used for their ability to produce sleep (major action), but they do not have the ability to relieve pain and cannot be used as narcotics. Thus, the sleep-producing effects of these two depressants overlap on a continuum of action chart (Figure 2.10). All the drugs that affect the central nervous system have similar properties. These actions were used by Dr. Robert W. Earle to produce the continuum shown in Figure 2.9. From the neutral area, specific points on the chart indicate the movement from the area of one group of drugs in terms of its major action into the area of another, more powerful group of drugs. The weaker groups of drugs are nearer the center; the most powerful drugs are at the two extremes.

If dosages are increased from minimal, to maximal, to toxic, to abusive, to lethal, any drug group listed in Figure 2.10 is able to produce the complete range of effects of stimulation on the one hand or depression on the other. This overstimulation or extreme depression produces the effects the drug abuser is seeking. Consequently, dosages used by drug abusers are far in excess of the dosages normally used in medical practice. A complete range of effects, produced by increased dosages, is presented in Figure 2.11. An example of the relationship between dosage and effects for one group, the stimulant drugs, is given in Figure 2.11.

Drugs can be arranged in such a continuum because the effects a person is seeking when abusing drugs are extremely similar. For exam-

NEUROLOGICAL ASPECTS OF ABUSIBLE DRUGS
48

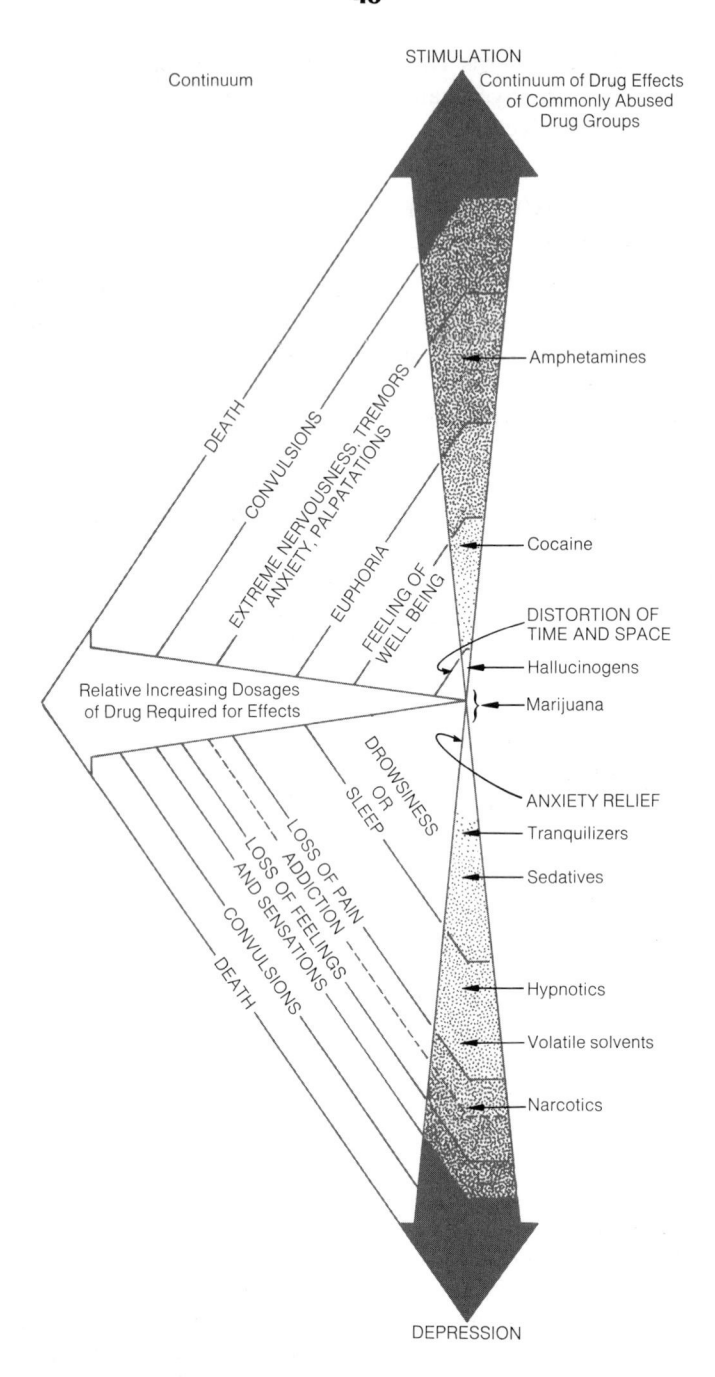

Figure 2.9. Spectrum and continuum of drug actions. Trinary relationship of increased dosages to specific drug groups to continuum of drug effects. As dosages are increased, the effects vary along a continuum until the dosage becomes lethal.

SPECTRUM AND CONTINUUM OF DRUG ACTIONS

Continuum of Drug Effects and Actions

- Death
- Stimulation ↑
- Convulsions
- Extreme nervousness, Tremors
- Anxiety, Palpatations
- Feeling of well-being, Euphoria
- Distortion of time and space
- Neutral area
- Anxiety relief
- Drowsiness
- Sleep
- Loss of pain
- Addiction
- Loss of feeling and sensations
- Convulsions
- Death
- Depression ↓

Overlapping of Drug Groups

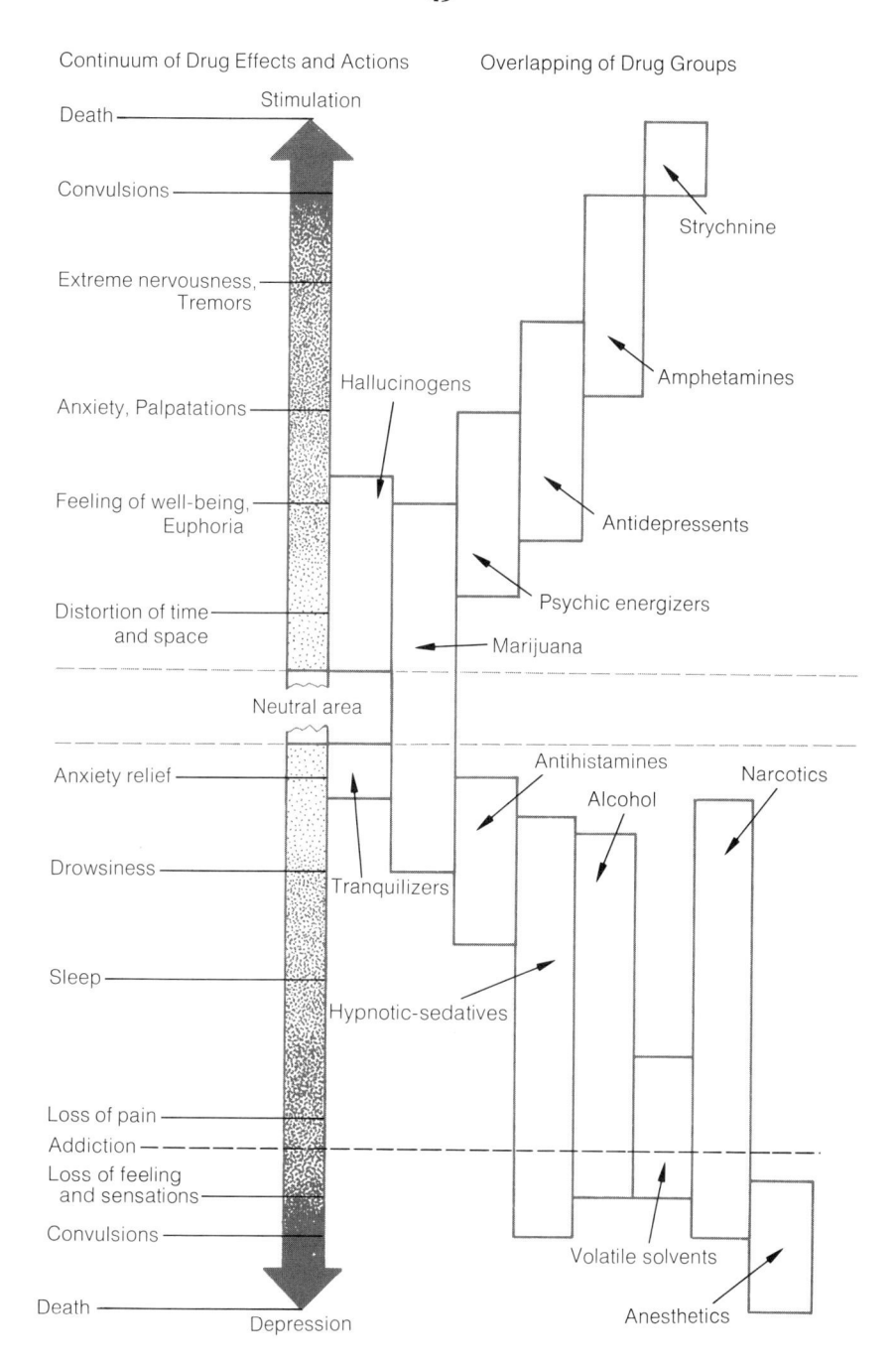

- Strychnine
- Amphetamines
- Hallucinogens
- Antidepressants
- Psychic energizers
- Marijuana
- Antihistamines
- Narcotics
- Alcohol
- Tranquilizers
- Hypnotic-sedatives
- Volatile solvents
- Anesthetics

Figure 2.10. Overlapping effects of abused drug groups.

NEUROLOGICAL ASPECTS OF ABUSIBLE DRUGS

50

Drug Effect						
DEATH		CONTINUUM OF DRUG EFFECTS				
						Stimulation
CONVULSIONS						
EXTREME NERVOUSNESS, TREMORS						
ANXIETY, PALPATATIONS						
FEELING OF WELL BEING, EUPHORIA						
DISTORTION OF TIME AND SPACE						
	Lethal Dosage	Abusive Dosage	Toxic Dosage	Maximal Dosage	Minimal Dosage	Neutral Area of Drug Continuum
			DOSAGE LEVELS			

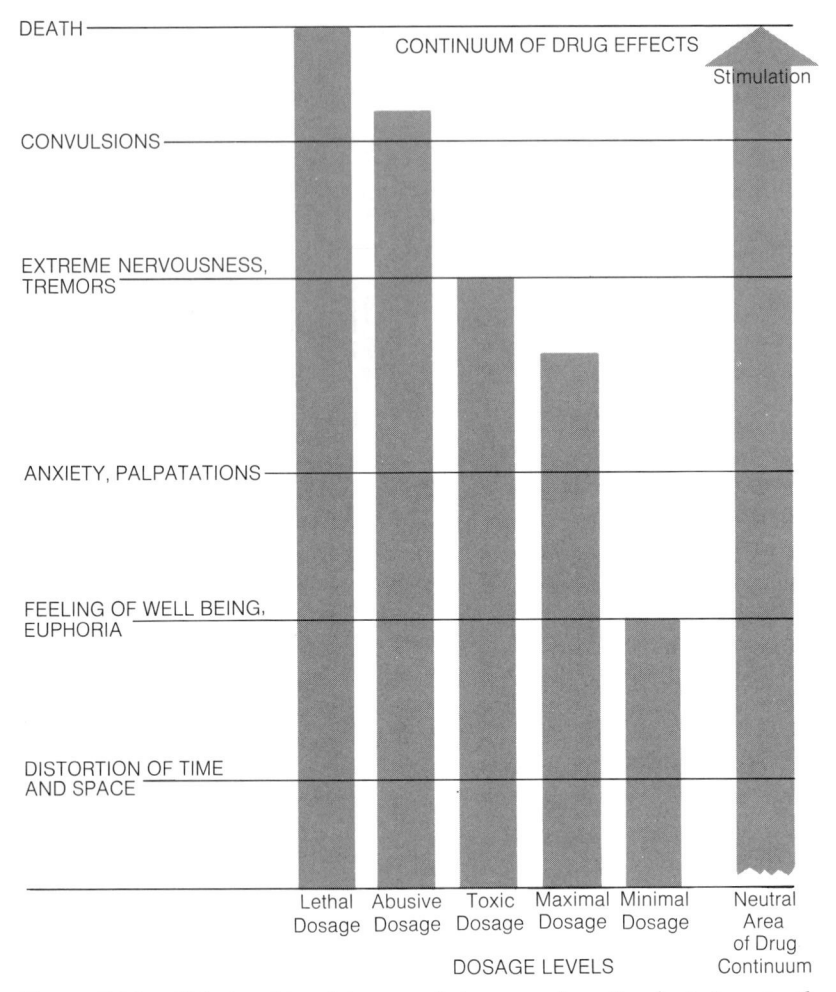

Figure 2.11. Relationship of increased dosages of a stimulant drug to the continuum of drug effects.

ple, any of these drugs will produce hallucinations at some dosage. This is why individuals, although preferring one drug over another, will abuse any drug within these groups if it becomes available. As the specific actions of a drug become more familiar and less spectacular, the individual may experiment with new ways to use the drug. He or she may progress from taking the drug orally to injecting it under the skin or into a muscle ("skin popping") to injecting it directly into a vein ("mainlining"). Or they may seek stronger and stronger drugs to produce more vivid effects, quicker actions, or longer-lasting experiences. Very few compulsive drug abusers are satisfied with experiences from just one drug. Most often, they will move toward the extremes of the continuum.

The more commonly abused drugs, such as marijuana, alcohol, or the barbiturates (sedatives or hypnotics), are close to the center of the chart when used in minimal dosages. The strong preference for these drugs lies in the ability of individuals to control the amount and consequently the relative effect of the drug. In a highly emotional state, users are able to increase the dosage or consume more of the drug to reach a more intense effect. If less emotionally disturbed at the time of abuse they are less concerned and often content with reduced consumption and effects. This ability to control drugs is offset when addiction levels (Figure 2.9) are reached with the depressant drugs. At addiction levels, the dosage must be increased regularly in order to keep the body from entering withdrawal, regardless of the individual's emotional state. This increasing dosage is required to keep the individual at normal or neutral levels. Thus, they have locked themselves into a constantly changing artificial continuum of effects. The minimal dosage must be used daily to maintain a neutral level, and then additional abusive dosages are needed to reach the levels of effects that will satisfy emotional states.

The chemistry and the pharmacological classification of marijuana now appear to be distinct from those of all other groups. However, the pharmacological action of marijuana has some similarities to properties of the stimulants and some likenesses to properties of the depressants. This is why it has been placed overlapping the neutral area of the continuum of drug actions (Figure 2.10).

When an individual abuses drugs that do not have the addictive properties of depressant drugs, an individual may not feel the need to progress from a relatively mild drug (marijuana), being better able to control the effects by controlling the amounts consumed. However, some who start on weaker drugs often progress to stronger ones because of the differences in the quality or intensity of effects.

The drugs within each group do differ slightly in their effects. This is why different drugs are used for different therapeutic purposes. These differing effects are regulated by the areas or routes of action within the central nervous system. Because of the complexity of the body's organization, drug action is often complex. This complexity of action at times makes it extremely difficult to place a group of drugs on a progressive-continuum chart. This is more of a problem with the stimulant drugs than with the depressant drugs. Also, at times the complexity of actions leads to apparent paradoxes. For example, if alcohol is a depressor of nerve function, why do people seem stimulated by a small amount of alcohol? The answer is that the brain contains a group of cells whose function is inhibition, cells that normally keep an individual from acting irresponsibly on every passing impulse. These inhibitory cells are more sensitive to alcohol and similar drugs (such as barbiturates and solvents) than the other brain cells are. As the alcohol concentration in the blood begins to rise, the inhibitory cells are depressed and cease to function properly, so that many impulses which would otherwise be suppressed are acted upon. Therefore, moderate amounts of alcohol, a depressant drug, can cause excitation by depressing an inhibitory function.

The following chapter presents the general classification or groupings of the drugs based on this continuum of actions associated with their use. The classes to be discussed are narcotics, volatile solvents, hypnotic-sedatives, tranquilizers, marijuana, hallucinogens, psychic energizers, antidepressants, and amphetamines.

SUMMARY

I. The main coordinator of body functions is the nervous system
   A. It consists of the *central nervous system* and the *peripheral nervous system*.
   B. The basic functional unit is the *neuron* or *nerve cell*.
      1. The junction between two neurons is a *synapse*.
      2. The junction between neuron and muscle is a *neuromuscular junction*.
   C. Transmission of nerve impulses across these junctions is done by production of neurotransmitter substances.
      1. The major transmitter is acetycholine (ACH). Such fibers are called *cholinergic fibers*.
      2. Other transmitters are classified as catecholamines. Such fibers are called *adrenergic fibers*.
   D. The central nervous system functions like a computer, and the peripheral nervous system like a telephone transmission line.
   E. Drugs influence specific neurons in the nervous system and either increase or decrease the activity of neurotransmitters.

II. How psychoactive drugs work
   A. A drug may act on the surface of cells or within cells.
      1. Drugs are selective and exert actions by becoming attached to a specialized chemical group or place on a cell membrane, called a *receptor site*.
      2. Receptor-binding of a drug to a specific cell is pictured as being like the fit of a key into a lock.
      3. Receptor-binding enables the drug to participate in a few steps of a cellular process thereby altering the cellular functioning.
   B. A drug that combines with the same receptor site as that of another drug but does not initiate the same drug actions is called an *antagonist*.

SUMMARY

C. Drug elimination within the body
   1. Drugs may be chemically inactivated, excreted from the body unchanged, or chemically transformed into less effective substances or waste products.
   2. Elimination of drugs from the body in either active or inactive forms occurs by way of the usual channels of excretion.

III. Drug-induced effects on the central nervous system
   A. Mood-modifiers, psychotropic, psychoactive, or psychotoxic substances bind to specific receptor cells in the central nervous system, producing their effects.
   B. The overall actions and effects of these drugs are those of either a stimulant or depressant.
      1. Central nervous system stimulants temporarily increase body function or nerve activity.
      2. Central nervous system depressants have the ability to decrease a body function or nerve activity temporarily.

IV. Structural components of the central nervous system.
   A. The *cerebrum* is a vast information-storage area. It is the site of conscious sensations and mental abilities such as memory, intelligence, imagination, creative thought, and recognition.
      1. Stimulation of the cerebrum increases conscious sensations, while depression decreases conscious sensations.
   B. The *thalamus* serves as a center for impulses to and from the cerebral cortex.
      1. Depression of various portions of the thalamus interrupt the free flow of impulses to the cerebral cortex.
      2. Stimulation increases the action of the cerebrum.
   C. The *hypothalamus* controls the basic life functions and is the mechanism for integrating all these basic functions with the emotions and behavior of the individual.
   D. The *cerebellum* is the center for muscle coordination, equilibrium, and muscle tone. Drugs that disturb it usually cause dizziness and loss of equilibrium.
   E. The *medulla oblongata* contains the vital centers (respiratory, vasomotor, and cardiac).
   F. The *spinal cord* is the center for the reflex activity and the transmission of impulses to and from the higher centers in the brain.

V. Functional components of the central nervous system
   A. *Reticular formation* extends from the spinal cord through to the cerebrum.
      1. Arousal reactions are the sensory activation of the reticular activating system of the reticular formation.

2. The reticular activating system, hypothalamus, and thalamus contribute greatly to the control of the degree of excitement and alertness a person feels.
B. The *limbic system* and its interconnections with the hypothalamus and thalamus are known as the *Papez circuit.*
   1. This circuit functions as a balancing system for the conscious experience of emotions and emotional expressions.
   2. Drugs that interrupt or cause this circuit to become unbalanced can change in individual's perception, expressions, and relative level of emotions.
C. Integrative reactions
   1. Any sensation coming into the brain causes the hypothalamus to send activating impulses back through a special division of the nervous system called the *autonomic nervous system.*
   2. Stimulation of the parasympathetic division is a stabilizing force.
   3. Stimulation of the sympathetic division increases energy output, increasing organ actions needed for increased effort.
   4. Nerves that transmit messages from the central nervous system through the autonomic nervous system consist of two neurons connected in a series.
      a. The first neuron is called the *preganglionic fiber,* and they release acetylcholine as the neurotransmitter and are cholinergic fibers.
      b. The second neuron is called the *postganglionic fibers,* and they release norepinephrine as the neurotransmitter and are adrenergic fibers.
   5. Drugs that affect preganglionic neurons are called *cholinergic drugs.* (Their older name was parasympathomimetic drugs).
   6. Drugs that affect postganglionic neurons are called *adrenergic drugs.* (Their older name was sympathomimetic drugs).
   7. Two antagonistic systems function within the hypothalamus through the autonomic nervous system.
      a. The ergotropic division functions through the sympathetic nervous system.
      b. The trophotropic system functions through the parasympathetic nervous system.
   8. Pain and pleasure—The hypothalamus and reticular formation regulate an individual's feelings of pain and pleasure.
   9. Rage—Stimulation of the pain areas also causes development of a classic emotional pattern called *rage.*
   10. Psychological and psychosomatic effects often result from

SUMMARY

hyperactivity of either the sympathetic or the parasympathetic system.

VI. Spectrum and continuum of drug actions
   A. The continuum of drug effects extends from over stimulation of the nervous system to death at one extreme and from severe depression of the nervous system to death at the other extreme.
      1. The weaker drugs are nearer the center; the most powerful drugs are at the two extremes.
      2. If dosages are increased, any drug group is able to produce the complete range of effects of stimulation on the one hand or depression on the other.
   B. Drugs can be arranged in such a continuum because the effects a person is seeking when abusing drugs are extremely similar.
   C. The following chapter (Chapter 3) presents the general classification or groupings of the drugs based on this continuum of actions associated with their use.

3
DRUGS COMMONLY ABUSED

Most people take drugs for medical reasons. People with headaches or allergies take aspirin or antihistamines. Individuals with peptic ulcers take antacids. When taken under the direction of a physician, antibiotics such as penicillin can be life saving in treating bacterial infections that were deadly 30 years ago. Sometimes a drug is taken as a substitute for chemicals the body may lack, for example, insulin. Drugs do preserve life and good health when properly used. But any drug strong enough to be effective can be harmful as well as beneficial if improperly used. The hypersensitivity of some individuals to drugs such as aspirin, penicillin, or antihistamines can produce untoward effects that may be severe enough to cause death, even when administered in small doses. Misuse of these drugs is usually a mistake or an accident. The drugs that are voluntarily abused are those that can modify the moods and behavior of an individual. As explained in Chapter 2, these drugs affect the central nervous system and are classified as psychoactive drugs. Some call these "recreational" drugs. Such drugs, regardless of their classification, either depress the nervous system ("downers"), or stimulate it ("uppers").

In their search for "recreational" drugs individuals have tried a limitless number of substances of unknown or unpredictable identity. Many of these substances constantly drift in and out of favor with drug users. Some individuals will always smoke all sorts of herbs and can be counted on to sample every over-the-counter drug they can get their hands on. Examples of these substances include cola and aspirin, catnip, nutmeg, morning-glory seeds, scotch broom, and most cough and cold

ANESTHETICS

remedies. This practice can be dangerous because often the effects of such substances are unknown.

This chapter will proceed up the continuum of drug actions (Figure 2.9) and describe each drug group from the more potent depressants (anesthetics) to the more potent stimulants (amphetamines). For the position and relative depression or stimulation abilities of each group, the reader should refer to Figure 2.10.

ANESTHETICS

Anesthetics are the most potent of the nervous system depressants. They can depress all cells of the body. Depending upon their actions, there are two types of anesthetics. *Local anesthetics* block nerve transmissions by depressing the fibers of neurons at any point in the nervous system where the drug is applied. *General anesthetics* act on the brain, producing *narcosis* (stupor or loss of consciousness) and a generalized loss of sensations. These drugs act by interfering with brain cells' ability to use energy.

During the stages of anesthesia individuals pass through a "stage of excitement." Some anesthetics have been used to produce this excitement. But such a practice is extremely dangerous because the individual can very easily lapse into the anesthesia stage of *medullary paralysis*—cessation of respiration and eventual death.

General Anesthetics

Early in this century *ether* and *chloroform* were used by a few as recreational drugs. Many of these people died in the process. As shown in Figure 2.10, other types of depressant drugs overlap the effects of anesthetics. Some of these are currently used as recreational drugs.

Sodium penthothal (sodium thiopental). This is a barbiturate which is commonly used as an anesthetic. It is abused, but not often. Because of its rapid action, a slight overdose can produce medullary paralysis and death.

Phencyclidine (PCP). The most widely abused anesthetic is *phencyclidine,* more commonly known as *PCP*. When sold as a white powder, it is called "angel dust," or in capsule or pill form, "peace pills." PCP dissolved in a liquid is most often sprayed on marijuana, parsley, oregano, or other plant leaves and smoked. It has been drunk or injected, but the dangers of an overdose by these methods are extremely high.

Phencyclidine was developed as an animal tranquilizer. Remember, when an animal is "tranquilized," it is "put-out" or anesthesized. This

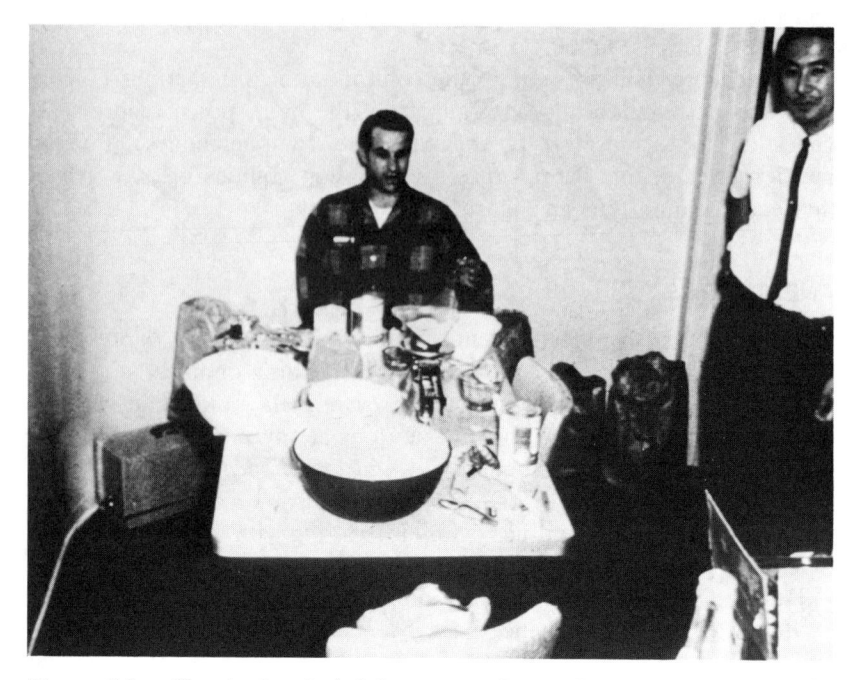

Figure 3.1. Illegal chemical laboratory. Illicit laboratories, such as this one, have sprung up across the United States and Canada for producing a number of drugs (LSD, PCP, & barbiturates).

drug is actually a potent general anesthetic, and very dangerous. Phencyclidine is not prescribed for humans because the range between an effective dosage and an overdose is too narrow for the general safety of the patient. Persons taking PCP hallucinate and feel "spaced out." They become aggressive, experience unsteady muscular coordination, and sometimes are unable to speak. Users feel the drug within 2 to 5 minutes and peak within 15 to 30 minutes. The effects continue for 4 to 6 hours, but often the user does not feel normal for 24 to 48 hours afterward. Heavy doses show all these symptoms plus eventual coma, and too often, death. The problem is that many PCP-related deaths go unrecognized. Accidents, homicides, suicides, and bizarre deaths are often suspected of being PCP-related.

Also, a large number of illicit laboratories producing PCP, such as the one shown in Figure 3.1, have sprung up across the United States and Canada. The chemicals used in its manufacture are extremely explosive, and a number of these laboratories have blown up, killing their operators.

NARCOTICS

The noun *narcotic* comes from a Greek word meaning "to benumb" and is defined as "having the power to produce sleep or drowsiness and to relieve pain." Narcotics consist mainly of the opiates (opium, morphine, heroin, and codeine) and the synthetic narcotics (such as Percodan, Dolophine, Demerol, or Darvon). (See Frontispiece, "Drugs Commonly Abused.")

Effects of Narcotics

Narcotics are medically used for their ability to produce an insensibility to pain *(analgesia)* and *sedation* (freeing the mind of anxiety and calming the emotions) without producing excessive drowsiness, muscular weakness, confusion, or less of consciousness, such as happens with general anesthetics. Narcotics cannot be used in place of an anesthetics. Narcotics cannot be used in place of an anesthetic, because in doses large enough to produce a loss of sensation (anesthesia) they depress the respiratory center in the brain (medullary paralysis) to a degree that can result in death. All narcotics have features in common, differing only in the degree to which actions or effects are produced.

In 1971, Avram Goldstein, Director of the Addiction Research Foundation at Stanford University, identified the *narcotics receptor* cells of the pain pathways in the brain, that is, the cells with which narcotics must combine in order to produce their effects. These receptor cells are found in the *central gray area* of the brain (Figure 3.2). Binding of the cells of this area in turn activates nerve cells in the lower areas of the cerebrum which inhibit the transmission of nerve impulses carrying pain signals to the brain from the body. Evidence also suggests that there may be receptor cells in other areas of the body, such as the nerves of the small intestine and the blood vessels near the skin. These receptor cells and others may explain the many side effects of narcotics, such as constriction of the pupils of the eyes, nausea and vomiting, reduction in body temperature, and dilation of the blood vessels near the skin. These areas may be activated by narcotics circulating in the blood stream, because individuals who remain quiet after taking narcotics have less nausea and vomiting than do individuals who increase circulation by physical activity.

In 1975, Dr. Avram Goldstein reported in *Life Science*[1] the isolation of a hormone from the pituitary gland which he called "pituitary opioid peptide" (POP). This hormone is very similar in chemical composi-

---

[1] B. M. Cox, K. E. Opheim, H. Teschemacher, and Avram Goldstein, "A Peptide-like Substance from Pituitary that Acts Like Morphine—Purification and Properties," *Life Science* 16(1975): 1777–1776.

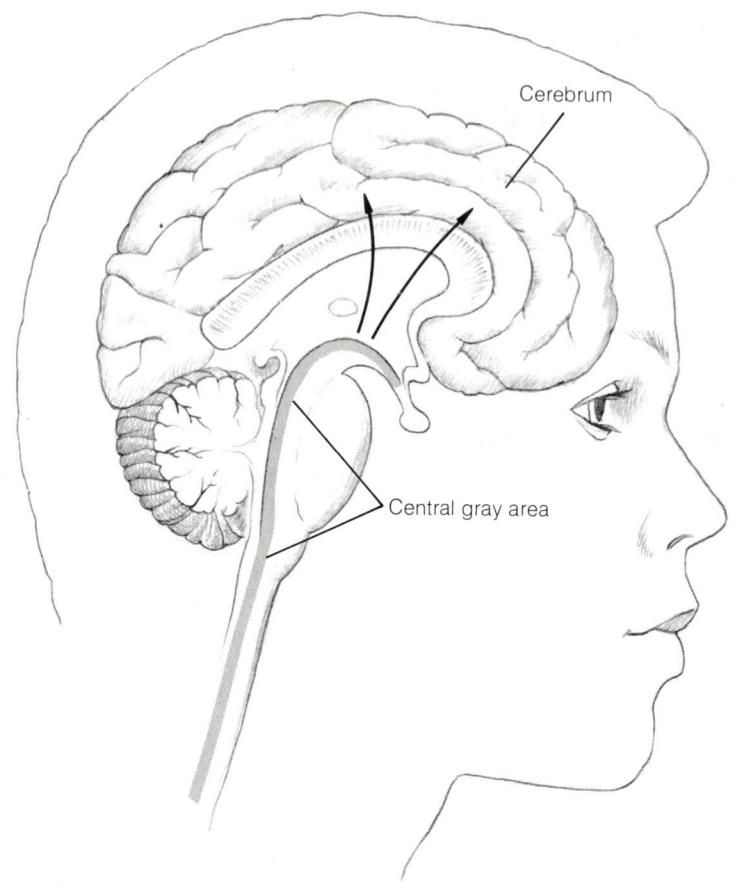

Figure 3.2. Central gray area. Binding of receptor cells of this area activates nerve cells in lower areas of the cerebrum which inhibit transmission of nerve impulses carrying pain signals to the cerebrum from the body.

tion to morphine and may be normally produced in our bodies by the middle lobe (pars intermedia) of the pituitary gland (Figure 3.3). Later the hormone was purified and named *endorphin* by psychopharmacologist Solomon H. Snyder of Johns Hopkins University in Baltimore. Also, Doctors John Hughes and Hans Kosterlitz of the University of Aberdeen, Scotland, isolated another group of morphinelike hormones in the brain, which they named *enkephalins*. Between 1975 and 1977, 6 enkephalins were isolated from the brains of mammals.

These endorphins and enkephalins seem to be released from the pituitary gland and bind to the narcotics-receptor cells in the pain (and other) pathways of the brain whenever we are under pain or stress. This allows us our normal "threshold ability" to withstand pain, anxiety, and stress. The amounts of endorphins and enkephalins produced,

NARCOTICS

and their ability to bind to receptor cells, give different individuals their differing thresholds to pain, anxiety, and stress.

When individuals are experiencing extreme or chronic pain, their endorphins and enkephalins may not be able to block all the discomfort. Thus, a narcotic can be given which will bind to any open receptor cells, thereby blocking the excess pain. As the pain subsides to a level the individual's endorphins or enkephalins can contain, the narcotic is no longer needed. Evidence suggests that the points of acupuncture analgesia (and the use of acupuncture in the treatment of narcotics addicts) causes the pituitary to secrete abnormally large amounts of endorphins and enkephalins. Also, electrostimulation through electrodes implanted in the receptor areas of individuals suffering from chronic or disabling pain can stop the pain. Each morning for a few minutes these individuals turn on their imbeded stimulators and are free of pain for hours. Research has found that the prolonged injection of endorphin,

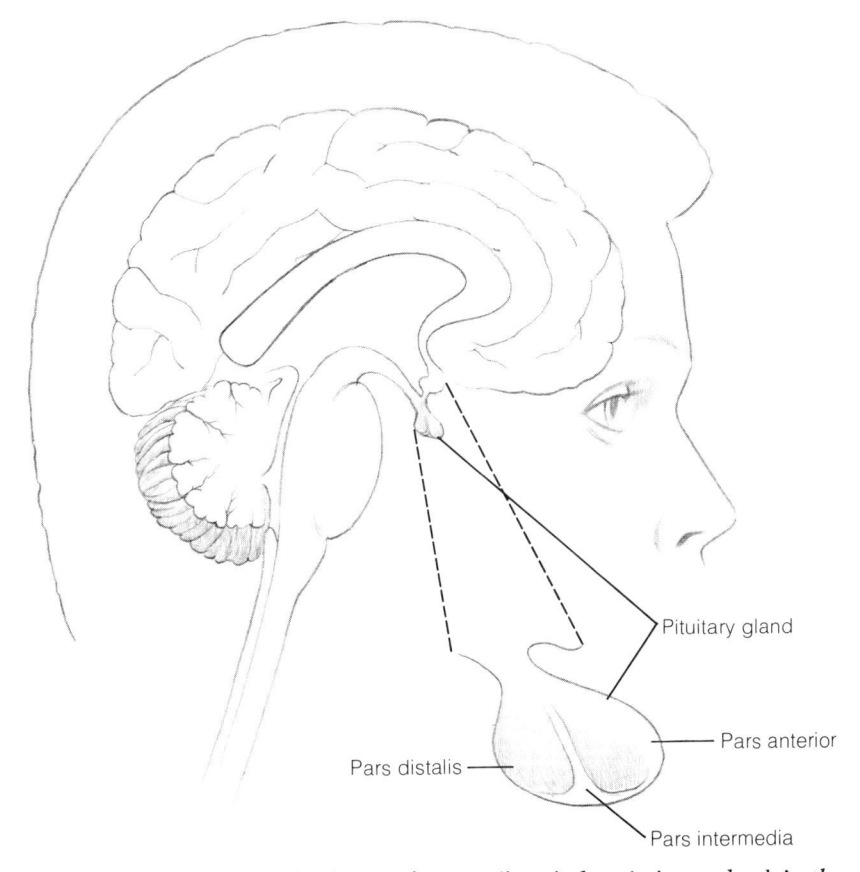

Figure 3.3. Pituitary gland. Pars intermedia *of the pituitary gland is the site of production of endorphin hormone.*

enkephalins, or electrostimulation produces tolerance and addiction just as the prolonged use of narcotics.

The use of narcotics and subsequent addiction may be greatly influenced by an individual's level of endorphin or enkephalins, by the amount of pain and stress in his or her environment, or by poor social, emotional, or cultural adjustment. Such individuals may habitually use narcotics to increase their ability to withstand the pain and stress of their environment. Or they may use narcotics to experience an elevation of mood, *euphoria* (an exaggerated sense of well-being and contentment), relief from fear and apprehension, and feelings of peace and tranquility. After experiencing these feelings for a short period, the state passes, and the individual becomes apathetic, with a slowing of both mental and physical activities terminating in sleep. Prolonged use of these drugs produces all the classical symptoms of drug addiction: Abrupt withdrawal causes *narcotic-solvent abstinence syndrome,* and a dependent user is clearly within the classification of an addict.

Effects of Abusive Dosages

Recognition of narcotics addicts, especially when they are regularly obtaining a daily dosage, is extremely difficult. Addicts will not reveal their condition, and experience makes them very adept at disguising it.

Abusive dosages of narcotics before much tolerance has developed produce constriction of the pupils of the eyes, as in Figure 3.4C. The pupils, constricted to a pinpoint condition, do not react to light; that is, the pupil will not constrict in strong light (even when light is flashed directly into the eyes) or dilate in weak light (as is the case when a bright light is quickly pulled away). This condition of the eyes is termed "frozen." After tolerance to a narcotic has developed to some degree, pupillary constriction may not be as pronounced, but the reduced ability to react to light remains. When addicted individuals have been without a dosage of narcotics for a period of time (from four to six hours or overnight), they begin to enter withdrawal, in which the pupils of the eyes become dilated but maintain the sluggish reaction to light.

Long-term addicts tend to be pale and emaciated and to suffer from severe constipation. Their appetites are poor, and they show little or no interest in sex. While under the influence of a narcotic, after the initial "flash" of euphoria, addicts become sleepy, lethargic, semistuporous, and dreamy (a condition referred to as "on the nod" or "doped"). While under the influence of drugs, narcotic addicts are not particularly dangerous. However, they suffer periods of extreme discomfort if their dosages are not taken three or four times daily, just to keep from going into withdrawal and to maintain a relatively neutral or normal condition. Thus, if deprived of a regular and necessary supply of a narcotic, they must steal to obtain money in order to maintain their habit, which

NARCOTICS
63

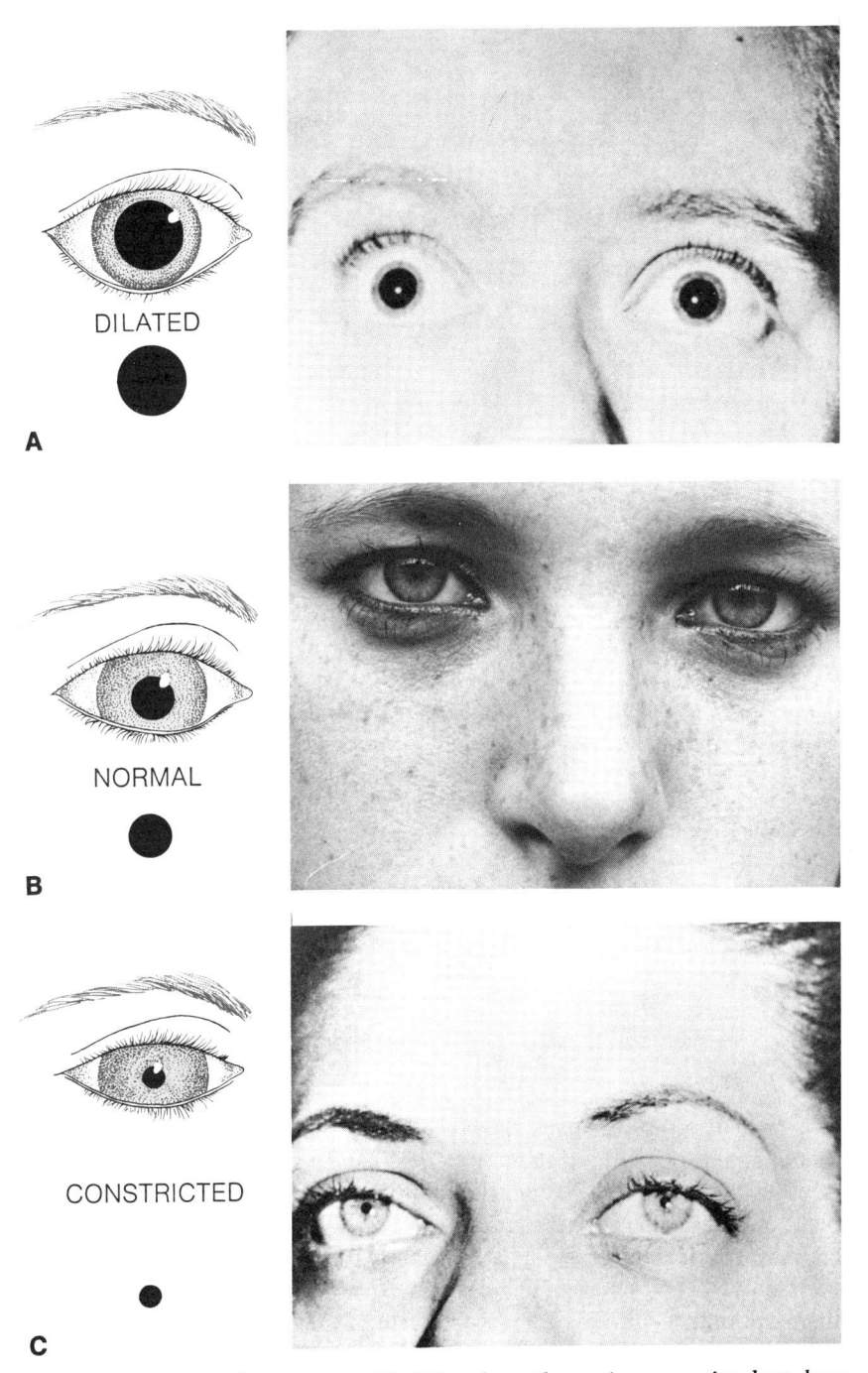

Figure 3.4. Pupil reactions: (A) *Dilated pupil reaction to a stimulant drug.* (B) *Normal pupil reaction in light of average intensity.* (C) *Contracted pupil reaction to a depressant drug.*

DRUGS COMMONLY ABUSED

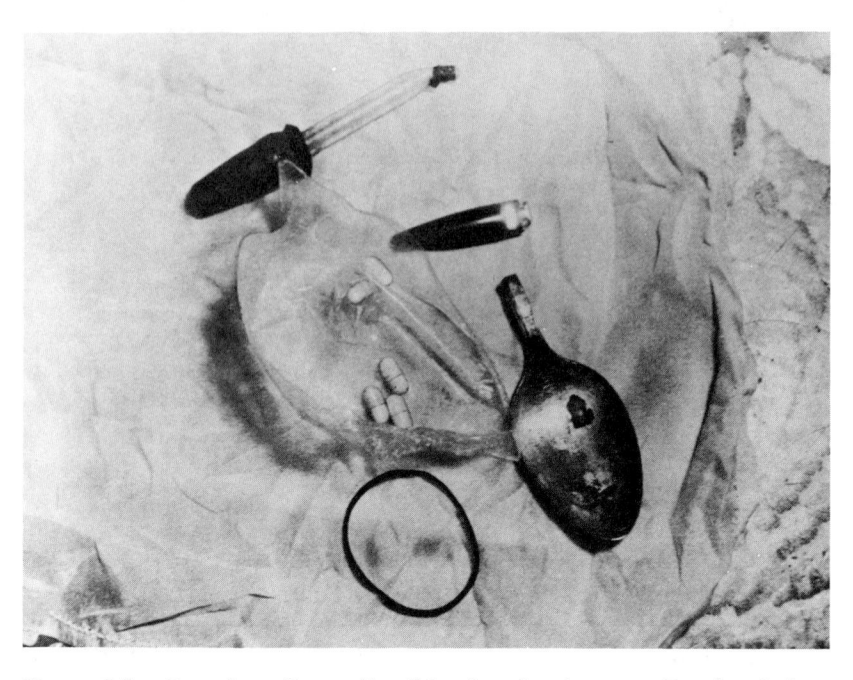

Figure 3.5. Paraphernalia or "outfit" of a heroin user. Powdered drug (heroin in this instance) is poured from the capsules into a spoon. Addition of a small amount of water and heating (by applying a match-flame under the spoon until the desired temperature, approximating body temperature, is reached) produces a solution. A small ball of cotton is placed in the bowl of the spoon and the mixture is drawn through this "filter" into an eye-dropper. The needle (inside the container in the photograph) is then affixed to the open end of the eye-dropper and secured by means of a gasket made of a scrap of paper, thread, rubber band, or tape. The injection is then made. (California State Bureau of Narcotic Enforcement.)

may cost up to $75 a day. New York City police officials have found that a large proportion of the city's crime can be traced to addicts. These offenses consist mainly of burglary, shoplifting, prostitution, and other petty offenses related to the addict's need for drug money.

Because of fear, haste, and a general disregard for themselves, addicts and narcotics users in general are not careful about sterilizing their equipment (Figure 3.5). As a result, infections and diseases such as syphilis and hepatitis may be passed on to other people who use the same equipment used by an infected individual. Overdose deaths occur from medullary paralysis depression of the respiratory center in the brain to a degree by that the individual stops breathing. If CPR (cardiopulmonary resuscitation) is administered until the person can be taken to an emergency hospital, they may be saved.

Administration of Narcotics

Narcotics may be administered by a physician either orally or by injection. A physician administering narcotics will usually inject them subcutaneously (injection under the skin) or intramuscularly (into a muscle). This type of injection, often used by addicts in the early stages of addiction, is known as a "skin pop" (shown in Figure 3.6). A "skin

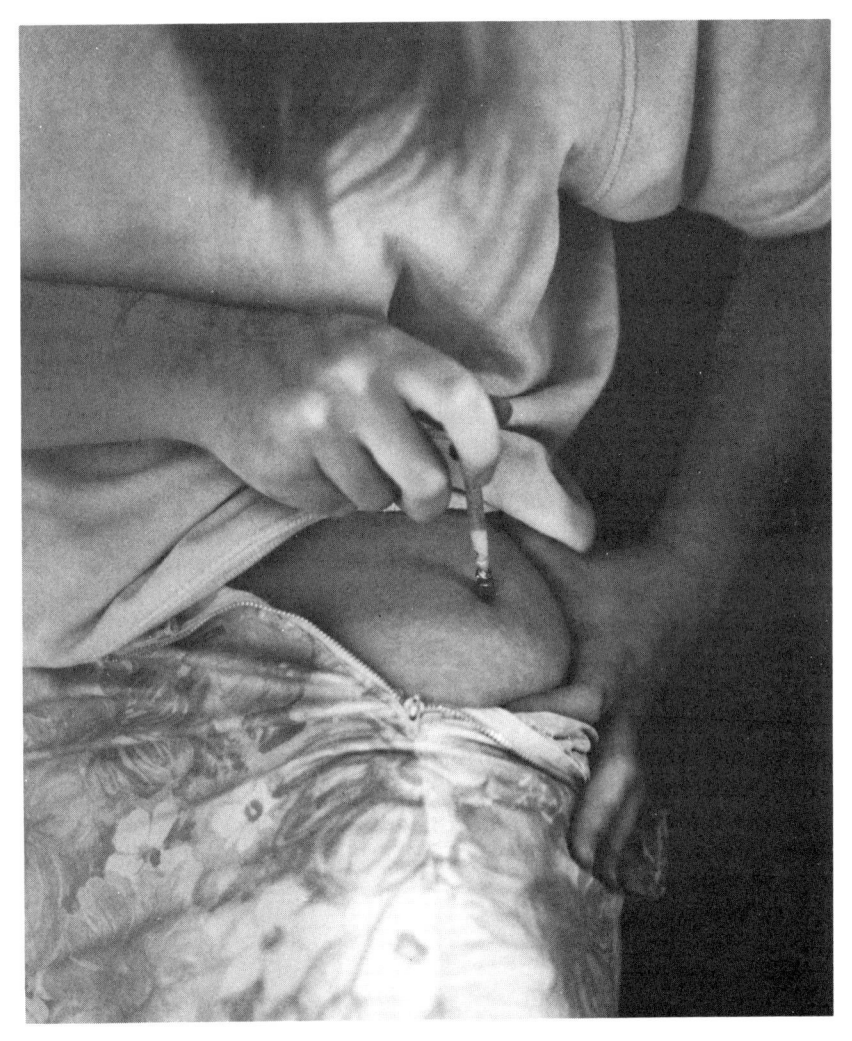

Figure 3.6. Injection under the skin (subcutaneous) or into a muscle (intramuscular).

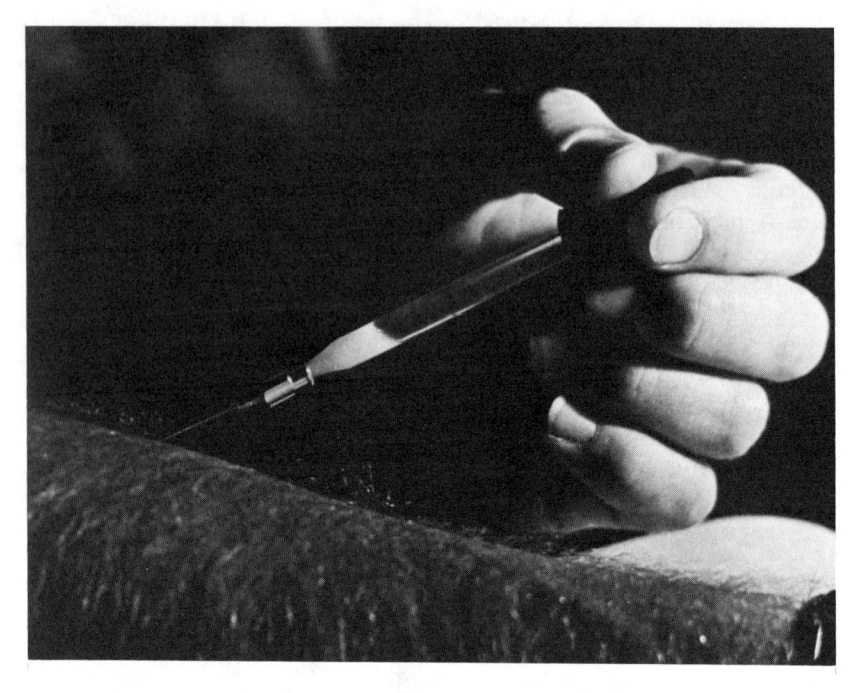

Figure 3.7. Injection directly into a vein (intravenous). At this point a drug user is referred to as a "mainliner." The area of the arm in which the injection is to be made is often rubbed to bring the vein closer to the surface; a tourniquet is wrapped around the upper arm and tightened, cutting off the circulation. This makes the vein very prominent and easy to see. The needle is then inserted into the vein with great care; if the vein is missed another insertion must be made. The user often draws blood into the syringe (eyedropper) to make certain the needle has penetrated the vein. The bulb of the eye-dropper is slowly squeezed and as the drug is injected into the vein, the tourniquet is loosened and the drug courses through the bloodstream. The effect is immediate.

pop" gives too slow an action for confirmed addicts, so they resort to taking intravenous injections, that is, injections directly into the blood stream through a vein (shown in Figure 3.7), which gives the desired effect immediately. This type of injection is known as a "mainliner" or "fix."

Drugs Classified as Narcotics

Opium. Opium is derived from the Oriental opium poppy *(Papaver somniferum)* shown in Figure 3.8. The fragile red, white, or purple flowers grow best in a hot, dry climate. They cannot endure heavy rains

NARCOTICS

when blooming. The drug opium is actually the air-dried juice obtained by cutting the unripe capsules, or pods, of the flower. Since each plant yields very little juice, large areas of poppies must be planted in order to produce any significant amount of opium. Consequently, they can be grown profitably only where the climate is favorable (hot and dry) and land and labor are cheap (as in India, Turkey, and in certain parts of Russia, China, Egypt, and Mexico).

In opium-producing areas, after the heat of the day has passed, the workers go through the fields and make incisions into the capsules of the poppy flowers. During the cool night, the milky white juice of the capsules oozes to the surface. As it comes in contact with the air, it oxidizes, thickens, and takes on a reddish brown color. The following morning, the workers again go among the plants; scrape off the now heavy, molasseslike fluid; and collect it on poppy leaves. The material

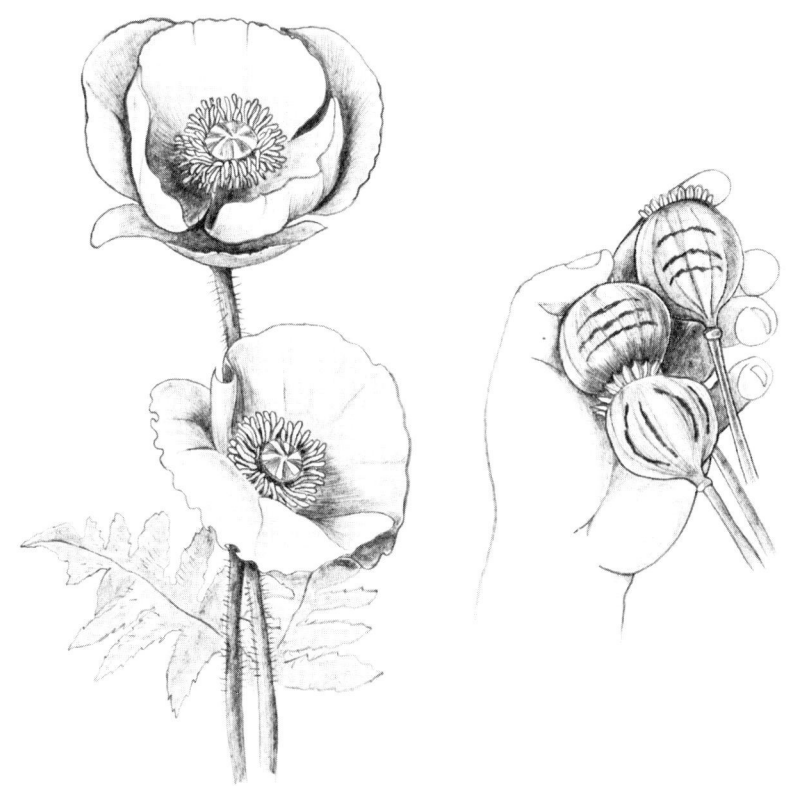

Figure 3.8. Oriental opium poppy (Papaver somniferum) *and capsules (pods) of the poppy flower. The pods show slashes for extraction of raw opium.*

gradually hardens, forming gumlike balls, almost black in color; this is opium in its raw state.

This sticky mass has a bitter taste and a heavy, sweet odor. When opium is being smoked, the odor is similar to that of wet, smoldering eucalyptus wood. Opium is generally smoked in a pipe or eaten. American drug users seldom use opium in its raw state, but they do use its derivatives.

Morphine and codeine are the most important narcotic substances obtainable directly from opium, which contains 10–16 percent morphine and 0.8–2.5 percent codeine.[2] The other opium derivatives, such as heroin, Dilaudid, thebaine, oxycodone (Percodan), and Prinadol, are obtained by chemical modifications of morphine and codeine or are produced synthetically.

Morphine.   Morphine is the chief derivative of opium. It is produced by the chemical refinement of opium. It can be pure white, light brown, or off-white; and it may be in the form of a cube, capsule, tablet, powder, or solution. On the illegal market, morphine that comes in a gelatin capsule is known as a "cap." The powder folded into a paper square is known as a "paper" or "package."

Morphine is about ten times stronger than opium. Therefore, its addictive attack upon the mental and physical condition of the user may be swift and strong. Unlike heroin, very little morphine is sold by peddlers. When morphine is unlawfully possessed, it usually has been stolen from a physician or pharmacist or obtained by means of forgery on a prescription form stolen from a physician. (Figure 5.1 shows the triplicate prescription formed used by physicians in prescribing narcotics such as morphine.)

Although morphine is usually administered by hypodermic injection, some drug users take it orally. Oral ingestion, called a "stomach habit," is not common because it necessitates a larger quantity of narcotic to obtain the desired euphoric effect and because the action is slower.

Heroin (diacetylmorphine).   Heroin is produced from morphine. It is from 20 to 25 times stronger than morphine, with twice morphine's addictive powers; thus, it is a very dangerous drug. In the pure state, it is a grayish brown powder. But because of its strength, the peddler is able to *adulterate* ("cut" or "step on") it many times before it is sold. Heroin may be cut with milk-sugar (lactose), mannite (a substance from the ash tree used as a mild laxative), procaine (a local anesthetic), or quinine. Often unsafe chemicals (such as LSD, amphetamines, PCP, or even strychnine) are used to adulterate heroin during

---

[2] M. Windholz, ed., *The Merck Index*, 9th ed. (Rahway, N.J.: Merck & Co., Inc., 1976), p. 6688.

the cutting process. Street heroin is often 4, 3, or even 1 percent or less in purity. Heroin is odorless and has a distinctly bitter taste.

The wholesale peddler sells heroin by the kilogram (2.2 pounds) down to 1-ounce or smaller plastic bags or in balloons—an easy way to store it. On the street the powder is folded into a paper square, known as a "paper" or "package," or in gelatin capsules, like morphine.

If the heroin is of high quality, such as is sold outside the United States, individuals may introduce the drug into their bodies through the nostrils by "sniffing" or "snorting" or through the lungs by mixing the heroin with cigarette tobacco and smoking it. These two routes of administration do not produce tolerance as quickly as injection does.

In the United States, a beginner may "snort;" but because of the poor quality of the heroin, he or she is soon forced to inject it to obtain the "kick" they are looking for. The addict begins by injecting it into the fleshy parts of the arm or body, but a rapidly mounting tolerance necessitates injection directly into the veins for the desired euphoric effects. At this point, the addict is referred to as a "mainliner."

A standard hypodermic needle attached to a syringe or common medicine dropper may be used. When a needle is not available, the drug may be forced into the body by opening a blood vessel with a pin or razor blade and administering the heroin directly with the medicine dropper. The equipment needed to inject heroin is termed an "outfit" (Figure 3.5). Generally, a tourniquet is attached to the upper arm to bring the main vein into prominence. With constant injection, the blood vessels break down, and scars form the veins, causing the addict to seek new areas for injection until the entire length of the arms is marked by needle punctures. Prolonged addiction results in the necessity of injecting between the fingers, in the legs and the neck, above the hairline, and even inside the mouth.

Heroin is about 4 times more toxic than morphine, and its action, at the toxic dosages used, has a more demoralizing effect. The danger of addiction to heroin is greater than that to any other narcotic because the body's tolerance (Figure 2.2) to the drug builds so rapidly. The addict requires increasingly larger dosages within a short time in order to secure the euphoric effects. Addicts prefer heroin to morphine and take morphine only when they cannot obtain heroin.

Legally, heroin is considered the most dangerous of all drugs, and because of its effects, the manufacture and sale of heroin in the United States have been prohibited by federal law since 1922 (see Chapter 5 on federal drug laws).

Codeine (methylmorphine). Codeine is milder than the other opiates in this section, and its power to induce sleep and relieve pain is relatively mild. It is widely used as an ingredient in cough medicines because of its effects on the cough center. It is manufactured either directly from opium or derived from morphine. It is an odorless, white crystal or crystalline powder and is taken orally as a tablet or in a solution or is injected. Generally, throughout the United States, preparations containing not more than one grain of codeine per ounce of fluid can be sold over the counter; but the pharmacist must record the name of the purchaser. In some states, preparations will bear a label with the words, "Contains Codeine (Opium derivation). WARNING—may be habit-forming. Do not give to children except upon advice of a physician."

Heroin users will sometimes resort to the use of codeine when deprived of their regular supply of heroin. But codeine is not widely abused because its euphoric effects on the reticular activating system in the brain stem are rather mild. A toxic dose of codeine often causes convulsions that are induced by the stimulating effect on the spinal cord, an effect which is able to override the weaker depressant effects on the higher centers of the brain.

Synthetic narcotics. Synthetic narcotics differ from the opiates in that they are made in the laboratory, starting with coal tar or petroleum products that have no narcotic properties before chemical conversion. Other narcotics are considered synthetic even though they are chemically derived from opiates as well as produced synthetically. Some of the more commonly known synthetic narcotics are Demerol, methadone, Mepergan, Percodan, Nucodan, Percobarb, Nalline, and Darvon.

Addiction to synthetic narcotics is less likely than addiction to morphine or heroin, but it is possible with all the synthetics. Withdrawal symptoms in the user are produced when any of these drugs is abruptly withdrawn. The abuse of synthetically produced narcotics, such as Demerol, is limited mainly to the medical and allied medical professions, where such drugs may be obtained without going to the illegal market.

VOLATILE LIQUIDS

Solvents

Some volatile (easily vaporized) chemicals, when inhaled, produce a state of intoxication that is characterized by drowsiness, dizziness, a slurring of speech, the loss of consciousness, and often, hallucinations. These chemicals are usually solvents (chemicals capable of dissolving something), which are contained in lighter fluid, paint thinner, cleaning fluid, gasoline, and model airplane glue.

Several dangerous solvents are used in the manufacture of airplane glue or cement. The most common are isoamyl acetate and ethyl acetate. Other dangerous solvents used are benzine, toluene (also called toluol), and carbon tetrachloride. Prolonged inhalation of any of these solvents

VOLATILE LIQUIDS
71

may cause death. The labels (Figure 3.9) on many of the fluids containing solvents have the warning: "Use only in a well-ventilated, open area." The hydrocarbons in gasoline (such as butane, hexane, and pentane) also cause solvent intoxication when inhaled. Probably the most commonly abused are the airplane cements containing the solvent toluene.

Solvents have a depressant action on the central nervous system that is similar to the action of barbiturates and narcotics. This is why they are placed between hypnotic–sedative drugs (barbiturates) and narcotics on the continuum of drugs (Figure 2.10).

The toxic effects of solvents have been carefully observed. They include irritation (producing cellular death) of the mucous membranes, skin, and the respiratory tract; cellular injury to the heart, liver, and kidneys; and bone marrow depression, which results in anemia (reduc-

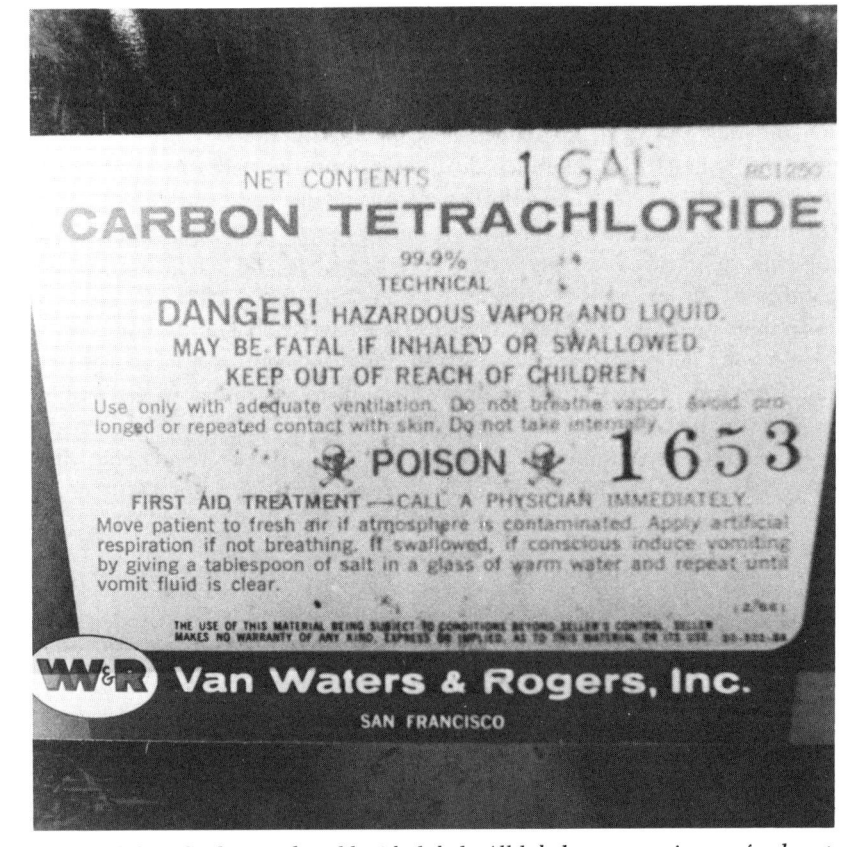

Figure 3.9. Carbon techrachloride label. All labels on containers of solvents include a warning against using them in a confined area.

tion in red blood cells), leukopenia (reduction in white blood cells), and thrombocytopenia (reduction in platelets in the blood). There have also been reports of brain tissue deterioration, acute liver damage, and kidney failures; many deaths have occurred as a result of repeated daily use of solvents.

In the beginning stage of glue sniffing, a few whiffs of the vapors will produce a "jag." But since tolerance to the solvents develops very rapidly, the habitual user must inhale the contents of many tubes of cement in order to experience the desired effects. By this time, many authorities feel he or she has acquired an addiction to the solvents.

The psychological effect is one of pleasantness, cheerfulness, euphoria, and excitement, closely simulating the early stages of alcohol intoxication. Abusers act drunk, exhibit disorientation, and have slurred speech. This period of drunkenness continues for 30–35 minutes after inhalation. Then they may become drowsy, lapse into a stupor, or become unconscious. They may remain unconscious for an hour or longer.

An addicted glue sniffer often has a characteristic unpleasant odor to the breath and excessive salivation. Such salivary secretions result from the solvent vapors' irritation of the mucous membranes of the nose and mouth. This irritation requires users to expectorate frequently. They also suffer from insomnia, nausea, and weight loss.

The strong drug dependence, the psychotoxic effects (during administration), the strong tolerance developed, and the necessity to increase the dosage characterize the solvents in abusive dosages as addicting drugs (Figures 2.9 and 2.10). However, since upon withdrawal from the solvents, a user experiences few severe withdrawal symptoms, some authorities do not classify solvents as addicting.

In their quest for volatile chemicals that have mood-modifying effects, some young people have tried to "sniff" anything that is in a pressurized can, from hair spray to the Freon gas used in refrigerators. When sniffed in a closed area, most of these spray-can substances displace the oxygen in the air, freeze the lung tissue, or coat the lungs with resins, and they have caused the deaths of hundreds of young people.

Amyl and Butyl Nitrite

Amyl nitrite is a volatile liquid that is sold in ampules covered with a nylon net so that they can be broken and inhaled. Consequently, they are known as "poppers." Butyl nitrite is sold as a "room odorizer" by names such as *Rush* and *Locker Room*. Nitrites dilate blood vessels and relieve the chest pains (angina pectoris) associated with some heart conditions. They also dilate the blood vessels in the brain, and some people use them at orgasm in an attempt to prolong the intensity of orgasmic sensations. But, nitrites also may cause a throbbing headache, facial flushing, nausea, or vomiting—almost guaranteed to reduce the beauty of the orgasm. Nitrites can also produce death if the person has a heart condition or is asthmatic.

HYPNOTIC-SEDATIVES

The hypnotic-sedative drugs are classified together because of their ability to depress the central nervous system into a condition resembling normal sleep. Drugs in this group are mainly general central nervous system depressants and are characterized by their broad range of suppression of brain functions. All these drugs resemble each other in this action regardless of their chemical differences.

The only difference between a hypnotic action and a sedative action is the degree of depression. The hypnotic action is a stronger depression of the central nervous system. When a drug of this group is given in a moderate-to-maximal therapeutic dosage, producing sleep soon after administration, it is known as a *hypnotic*. When it is given in reduced (minimal) dosages, even several times a day, in order to reduce excitement, it is called a *sedative*. With increasing dosages, all drugs in this group tend to produce a continuum of effects from tranquilization and sedation (the allaying of excitement or quieting) to the loss of psychomotor and intellectual efficiency, to artificial sleep, and then to coma and death. When abused (in toxic dosages), they greatly reduce anxiety and produce a mild euphoria as a relief for people who tend to live in an uncomfortable, painful, or hyperaroused state.

Barbiturates

The most widely used hypnotic-sedative drugs are the barbiturates. They were first produced in 1864 by combining urea (an animal waste product) with malonic acid (derived from an acid in apples). The compound obtained, a new synthetic, was named "barbituric acid." Since then, chemists have produced a great variety of derivatives of barbituric acid. More than 2500 compounds have been synthesized. The first derivatives used as hypnotic-sedatives were barbital (Veronal) and phenobarbital (sold under the trade name Luminal). About 50 barbiturate drugs, with various *latencies* (period of time between administration and effect) and *lengths of action* (period of time when drug is effective), have been produced and put on the market. On the illegal market, barbiturate drugs are known as "goofballs," "downers," or "sleeping pills" or by their colors. The individual names of some of the more commonly used and abused barbiturates are shown in Figure 3.10 and in the Frontispiece, "Drugs Commonly Abused."

Actions of barbiturates. In low and moderate dosages, it appears that the barbiturates interfere with oxygen consumption and the mecha-

DRUGS COMMONLY ABUSED

COLOR AND SHAPE OF CAPSULE	TRADE (AND GENERIC) NAME	STREET NAME	CHIEF MEDICAL USE	DURATION OF ACTION
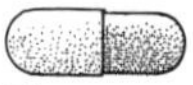 Blue green	Amytal (amobarbital)	"Blue dragons"	Sedative and hypnotic	Intermediate
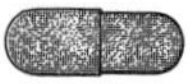 Red	Seconal (secobarbital)	"Reds" "Red birds" "Red devils"	Hypnotic	Short-acting
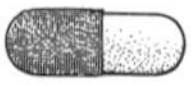 Red blue	Tuinal (combination of amobarbital and secobarbital)	"Rainbows"	Hypnotic	Moderately long-lasting
 Yellow	Nembutal (pentobarbital)	"Yellow jackets"	Hypnotic, sedative, and anticonvulsant	Short-acting
 Green	Luminal (phenobarbital)	"Purple hearts" "Barbs"	Hypnotic	Long-acting

Figure 3.10. About fifty of the several hundred barbituric compounds synthesized and tested are satisfactory for medical use. The barbiturates, which act on specific areas of the brain and induce sleep promptly, are effective in treating conditions of anxiety, muscular twitching, tremors, and convulsions. These compounds differ chiefly in the duration of their effects.

nisms by which energy is derived, stored, and utilized within the cells of the brain. These also become tolerant of barbiturates in time, requiring further dosage increases. This depression of brain cell function results in a general depression of function of most, if not all, areas of the central nervous system. Such depression of cellular activity accounts for the dosage continuum that produces suppression of the thinking, doing, and feeling abilities of the person taking barbiturates leading toward sleep, anesthesia, coma, and death (see Figure 2.9).

In higher dosages, barbiturates depress the synapses of neurons bringing information into the reticular formation, thus reducing an individual's wakefulness and alertness. They also depress the ability to stimulate both pre- and postganglionic synapses (by inhibiting the neurotransmitter substances) and motor areas of the cerebral cortex.

However, it appears that although many barbiturates act at all levels of the nervous system, some are very selective in their sites of actions. For example, certain barbiturates (phenobarbital, metharbital, and mephobarbital) work only on the cerebral motor areas, which enables these drugs to selectively prevent epileptic seizures.

At higher dosages, central nervous system depression lets barbiturates act as an anesthetic (a drug used to produce a loss of feeling and sensations). Not only does the individual become unconscious but the spinal reflexes are depressed to the extent that the muscles become relaxed and manageable for surgery.

However, at extremely high (toxic, abusive) dosages, in the dependent abuser (whose liver is active and able to detoxify these substances quickly), barbiturates produce a state of hyperactivity and excitement in the reticular activating system before the general suppression of the central nervous system sets in. This increased activity seems to be in response to depression of the brain's inhibitory systems. This inhibition-relieving action of barbiturates is responsible for a number of side actions associated with barbiturates: euphoria, excitement, and release of anxiety. Also, antagonistic actions upon the brain are what produce the "truth serum" recovery of memories; the antisocial, mood-modification, and behavioral changes of the "goofball" addict; or the occasional excited, sleepless night spent by some people who use sleeping pills night after night to go to sleep. The tolerance of the cells of the brain and spinal cord to the effects of barbiturates develops very quickly. This requires chronic users (or addicts, by this time), who are seeking this antagonistic action, excitement, and euphoria, to increase the dosage often to a lethal level without even suspecting the danger of the massive amounts of the drug they are using.

High dosages depress the respiratory and vasomotor centers (see Figure 2.3). Death from a lethal overdose is caused by respiratory failure accompanied by extremely low blood pressure. However, all depression of the nervous system is reversed following the inactivation or removal of the drug. Barbiturates are removed from the body mainly by the kidney. They are also altered by the liver to an inactive form. In dependent users or addicts, the liver becomes quite active and is able to inactivate (detoxify) larger and larger dosages faster and faster, which is one form of tolerance, with the result that increasing dosages must be used.

Effects of barbiturates. The barbiturates seem to be the second most popular suicide poison (carbon monoxide from automobile exhaust is first) in the United States. These pills account for the majority of drug poisoning, and most of these are considered suicide attempts. Actually, some of these deaths, although self-inflicted, are not suicides but accidents. Many may be overdoses by addicts. Accidental deaths may occur when a person has taken a moderate (prescribed) dosage to go to sleep and then, in a half-asleep and confused condition due to the effects of the dosage, takes another, lethal one. Physicians constantly warn barbiturate-using (sleeping pill) patients not to keep their bottles of tablets near the bed, where they may stretch out a hand to take more pills while in a confused or even comatose state of mind. These toxic and lethal effects of barbiturates are unpredictable. For some individuals, a comparatively moderate or small dosage may cause the effects of abuse and be dangerous. With these people, a relatively small dosage could also be lethal.

Long before the first tranquilizers were discovered, the barbiturates were being used as tranquilizers. Indeed, had phenobarbital been introduced 15 years ago instead of a half a century ago, it might have evoked the same burst of medical enthusiasm as Miltown (a minor tranquilizer) and its fashionable contemporaries. Still, use of the barbiturates is spectacular in terms of total production and consumption; each year people in the United States take an estimated 3–4 billion doses of barbiturates prescribed by their physicians.

Barbiturates are usually taken orally (barbiturate abusers use the term "dropped"). However, habitual users and addicts have been known to dissolve the compounds and inject them hypodermically. Sometimes they are "dropped" with alcohol and sometimes with Benzedrine or Dexedrine (amphetamines), which are central nervous system stimulants, in order to overcome the depressing effects of the barbiturates and to increase their antagonistic actions. This use of a stimulant drug to antagonize the depressant drug is dangerous, but the practice, widely used by young people, of consuming barbiturates and alcohol is even more dangerous and often results in death. Such a combination interferes with the body's normal disposal of both alcohol and barbiturates through the liver, causing a toxic or lethal level of each to be reached very quickly. Also, the two drugs working together have a synergistic effect; that is, the total depressant effect is far greater than the sum of their individual effects. Consequently, the consumption of even small amounts of barbiturates and alcohol in combination can be dangerous and may result in death.

A person under the influence of barbiturates acts like someone who has had enough alcohol to show signs of intoxication. How much of the drug is necessary to produce the degree of intoxication observed depends mostly on how accustomed to the use of the drug the person is. Dependent users keep taking more and more, and in time they reach amounts that would kill anybody who had not grown accustomed to the drug gradually. Whenever a person acts as if he or she has had a little or a great deal to drink but there is no odor of alcohol about them, it is possible that they have been abusing barbiturates. Sometimes when barbiturates and alcohol are taken together, they produce what looks like an ordinary "drunk," but the "drunk" takes much longer to sober up. People who get intoxicated on barbiturates follow the same course as people who take a drink of alcohol and keep on drinking until they pass out. However, barbiturates are more dangerous than alcohol because they are not vomited, and all the drug that is taken will be absorbed unless the stomach is pumped.

A small abusive dose (the equivalent of a large therapeutic dose) makes people feel relaxed, sociable, and good-humored; but they lose alertness and are very slow to react. After taking more, they become sluggish, gloomy, antisocial, maybe even quarrelsome. The tongue becomes thick; they stagger about for a while; and then gradually they slump into a deep sleep. Or if they have had a large amount of the drug, they may suddenly collapse into a coma. Users may die in the coma unless they receive medical attention promptly. Even when there is no apparent sign of life from a person in a coma, a doctor may be able to revive them.

Those who become addicted to barbiturates must have the drugs in order to prevent going into withdrawal (*alcohol-sedative withdrawal syndrome*). Without the drug, they have seizures that resemble epileptic convulsions. Sudden withdrawal of an addict from large abusive dosages of barbiturates without medical attention often results in death.

Alcohol and Alcohol Derivatives

For thousands of years, the only sleep-producing drugs that were available were alcohol, opium (now classified as a narcotic), and belladonna (a highly toxic stimulant drug extracted from *Atropa belladonna,* the deadly nightshade, a plant found in Europe and Asia).

The oldest hypnotic-sedatives used in modern medicine were alcohol and its major derivatives. The first drug derived from alcohol was *chloral hydrate,* used as a sleep-inducing drug since 1869. The next important hypnotic was another kind of alcohol derivative called *paraldehyde,* introduced into medicine in 1882.

Alcohol, when consumed in abusive amounts, is truly addictive; and the sudden withdrawal of alcohol from an addicted person produces serious disturbances. These may vary from craving for alcohol, anxiety, and tremors to full-blown delirium tremens (see Chapter 7). Although some psychiatrists believe that delirium tremens is a form of acute toxic psychosis, most experts in the field of alcoholism (alcohol addiction) regard it as *alcohol-sedative withdrawal syndrome.*

The abuse of alcohol is a problem of such magnitude that two complete chapters (Chapters 6 and 7) of this book are devoted to it.

Although chloral hydrate (see Frontispiece, "Drugs Commonly Abused") and paraldehyde are effective sleep-inducing drugs, the medical use of both has decreased, largely because the barbiturates are much more convenient to administer. Both chloral hydrate and paraldehyde have such disagreeable tastes and odors that they must be taken in special solutions to disguise these unpleasant properties.

Chloral hydrate is usually taken orally. Because it produces rapid and refreshing sleep, it has been called, and used as, "knockout drops." The lethal dose of this drug is highly variable among different individuals; thus, it is extremely dangerous to administer. Its effects are greatly increased (synergistic action) when there is the simultaneous use of alcohol, a condition that has been responsible for many accidental overdoses and deaths.

Paraldehyde is a liquid with a highly disagreeable odor. Oral administration, usually over shaved ice or in some cold drink, induces rapid sleep in most persons. Paraldehyde is seldom used as an ordinary sleeping drug because of its taste and its odor, which is noticeable to others for many hours following ingestion. It is given to hospitalized patients in the management of delirium tremens, withdrawal illness, and convulsions.

Abuse of these two alcohol derivatives is very uncommon because of their unpleasant tastes and odors.

Bromides

The bromides exert a sedative effect on the central nervous system. They were widely utilized in medicine and by the general public in the past. But with a growing recognition of the dangers of chronic bromide intoxication and of the cumulative action of the drugs and with the development of much more effective sedatives (the barbiturates), the modern physician finds few uses for bromides. They are still important to a drug study, however, because of the problems of bromide intoxication and poisoning.

The administration of sodium bromide or other bromide salts produces sedation, drowsiness, and sleep. Chronic administration of bromides tends to produce mental depression, confusion, and lethargy. Many individuals with bromide intoxication may be suspected of suffering from emotional disorders; in fact, some individuals have been admitted to mental hospitals because the toxic nature of the symptoms was not recognized. In bromide intoxication, various skin lesions, intestinal disturbances, and destruction of the membranes of the eyes and respiratory passages are common.

Bromides, although capable of inducing psychic dependence, are rarely abused because they do not produce euphoria. These drugs are highly dangerous because of their extreme toxicity.

Other Nonbarbiturate Hypnotic-Sedatives

There are other hypnotic-sedative drugs that produce actions ranging from anesthetics to tranquilizers (see Figure 2.10). Some drug users will try anything, and as these become available they are being abused more and more.

Ethchlorvynol (Placidyl).   Ethchlorvynol is a colorless-to-yellow liquid. It is a mild sedative which acts within 15–30 minutes and lasts for about 5 hours. Side effects from an overdose include headache, dizziness, mental confusion, nausea, and vomiting.

Ethinamate (Valmid).   Ethinamate is a mild sedative which is related to tranquilizers. It acts within 15–25 minutes and is effective for about 4 hours. It causes addiction, but does not seem to produce tolerance.

Glutethimide (Doriden or "Ciba").   Glutethimide is a hypnotic and sedative that depresses the central nervous system and produces effects similar to those of the shorter-acting barbiturates. It acts within 15–30 minutes, and its effects last 4–8 hours. It causes tolerance and produces addiction. Side effects are similar to those of the barbiturates, and deaths have occurred from overdoses.

Methyprylon (Noludar).   Methyprylon is similar to a barbiture. The onset of action is about 30 minutes, and the duration of action is about 7 hours. Others in this group of drugs include a new hypnotic related to the minor tranquilizers, Dalmane, Ditran, JB, and LBJ.

ANTIHISTAMINES

Histamine is a material that occurs naturally and is found in most cells of the body. It has very marked actions in very small concentrations. These actions appear to be almost always harmful. It produces a direct stimulation of certain smooth muscles and is a powerful vasodilator in the capillary beds. Upon release of histamine in humans, a noticeable dilation of the arterioles and capillaries is seen. This produces a flushing of the skin, a rise in skin temperature, and a fall in blood pressure. There is also vasodilation in the meningeal blood vessels, accompanied by an increase in intracranial pressure. The increased pressure often produces headache.

Antihistamines act by preventing the actions of histamine. They are quite selective in their abilities to block histaminic actions. Some of the more common antihistamines and their actions are listed in Table 3.1.

These drugs find their greatest uses in the relief of allergic responses, bronchial spasm in asthma, wheals (smooth, elevated, red or pale areas on the skin, a skin reaction to an allergy), and itching of the skin following an allergic response.

One peculiar action of some antihistamines is their ability to relieve or abolish the symptoms of motion sickness. A number of the compounds sold for this purpose are moderate to large dosages of antihistamines, such as dimenhydrinate (Dramamine), which is a chemically changed compound of diphenhydramine (Benadryl).

TABLE 3.1
Commonly Used Antihistamines

Official Name	Brand Name	Toxicity and Side Effects	Remarks
Brompheniramine	Dimetane	Mild drowsiness	
Carbinoxamine	Clistin	Drowsiness, dizziness, dryness of mouth	
Chlorcyclizine	Perazyl	Drowsiness	
Chlorpheniramine	Chlor-Trimeton	Mild drowsiness	
Clemizole	Allercur	Drowsiness, nausea, insomnia, excitation	
Cyclizine hydrochloride	Marezine	Drowsiness, nervousness, dizziness	May cause fetal deformities; contraindicated during cyclopropane anesthesia
Cyproheptadine hydrochloride	Periactin	Drowsiness	Contraindicated in glaucoma or urinary retention
Dexbrompheniramine	Disomer	Mild drowsiness	Dextro isomer of bropheniramine
Dexchlorpheniramine	Polaramine	Mild drowsiness	Dextro isomer of chlorpheniramine
Dimenhydrinate	Dramamine	Drowsiness	
Dimethindene maleate	Forhistal	Sedation, drowsiness	
Diphendydramine	Benadryl	Drowsiness	Also has some antispasmodic action; may help in bronchial asthma
Diphenylpyraline	Diafen	Drowsiness, dryness of mouth	
Meclizine hydrochloride	Bonamine, Bonine	Drowsiness, dryness of mouth, blurred vision	May cause fetal deformities
Methapyrilene	Histadyl	Drowsiness, vertigo, anorexia	Sedative effect
Methdilazine	Tacaryl	Drowsiness, dizziness, nausea	Phenothiazine derivative
Phenindamine tartrate	Thephorin	Insomnia, nausea, vomiting, kidney damage	

TABLE 3.1 *(Continued)*

Official Name	Brand Name	Toxicity and Side Effects	Remarks
Pheniramine	Trimeton	Drowsiness	Seldom used
Promethazine	Phenergan	Potentiates sedative and narcotic drugs, drowsiness	Phenothiazine derivative, like chlorpromazine; therefore, also acts as a tranquilizer and antiemetic
Pyrrobutamine	Pyronil	Not yet clear	Long-acting
Rotaxamine	Twiston	Drowsiness, dizziness, dryness of mouth	Levo isomer of carbinoxamine
Trimeprazine	Temaril	Drowsiness, dizziness, nausea, agranulocytosis	Phenothiazine derivative
Tripelennamine	Pyribenzamine	Gastrointestinal upset	Less likely to cause drowsiness than most antihistamines
Triprolidine	Actidil	Drowsiness, dizziness, dryness of mouth, nausea	

Source: Solomon Garb, Betty Jean Crim, and Garf Thomas, *Pharmacology and Patient Care,* 3rd ed. (New York: Springer, 1970), Table 55.2, pp. 391–392.

The antihistamines are placed on the continuum of drug actions (Figure 2.10) and are included in this discussion of hypnotic-sedative drugs because of some of their side effects. Many of the antihistamines produce drowsiness that can progress into sedation and produce sleep when used in increasing dosages. When this property was first noted, it was considered a minor side effect which, although undesirable, could be tolerated by most individuals. However, today, many drug manufacturers have utilized this side effect of drowsiness as the major active ingredient in most of the over-the-counter tension relievers (such as Cope) and sleep-producing pills (such as Sleep-eze). This use is possible because most antihistamines and their compounds can be purchased without a prescription.

There is an increasing usage of these compounds, and it should be noted that in some states, such as California, it is a misdemeanor to drive an automobile while under the influence of an antihistaminic drug because of the drowsiness the drug produces. Also, if a driver under the influence of such a drug is in an auto accident in which there is bodily harm to someone, he or she may be charged with a felony.

TRANQUILIZERS

The tranquilizers (see Frontispiece, "Drugs Commonly Abused") are a group of drugs able to relieve or prevent uncomfortable emotional feelings by reducing levels of anxiety. They relieve tension and apprehension and promote a state of calm and relaxation.

As shown in Figure 2.10, there are overlapping areas of effects with other drug groups. Like the barbiturates, many tranquilizers have sedative and hypnotic effects. And the barbiturates, in turn, have some tranquilizing actions. Even the narcotic drugs have tranquilizing actions. The tranquilizers are considerably less addicting (see position on Figure 2.9) than the narcotics, but addiction can occur if large enough dosages are used over an extended period of time.

The dramatic effects of the minor tranquilizers on the general public and of the major tranquilizers in the treatment of violent, overactive, psychotic individuals have, in some instances, led to exaggerated expectations regarding their role in the treatment of the mentally ill. Tranquilizers do not cure mental illness, but they do reverse many of the symptoms of psychosis. They make the management of the mentally ill who are taking the drugs easier and put the patient in a more desirable state of mind for psychotherapy.

The Major Tranquilizers

The major tranquilizers are more suited to management of the severely mentally ill and are more toxic. The introduction of these drugs has had a great impact on psychiatry. They are not tranquilizers in the sense that they make a person feel more tranquil. These drugs are actually powerful *antipsychotic agents*. Because of chemical similarities to other drugs, the major tranquilizers probably depress or block the actions of neuro-transmitters, such as norepinephrine and acetylcholine, that function in and around synaptic connections in the reticular formation; but no definite chemical interpretation has been established. Some of them (reserpine and phenothiazine derivatives such as chlorpromazine) are thought to be able to suppress brain centers, such as the hypothalamus, associated with the sympathetic nervous system, letting the parasympathetic system take control. Their effects on normal individuals are varied, but in general they range from deep sleep to a general feeling of depression. These are dangerous drugs; their use is not recommended except when a person has or is in danger of developing a severe mental disturbance (psychosis). The use of the major tranquilizers for nervousness, even in the low dosages advertised by some drug firms, is dangerous and not justified.

The first major tranquilizer reported as helpful in management of psychosis was a rauwolfia alkaloid, *reserpine,* extracted from a large climbing or twining shrub (*Rauwolfia serpentina*) that grows in India and various tropical regions of the world. In its raw root form, reserpine had been used in the treatment of hypertension (high blood pressure) and psychosis for many years in India before it was used in the United States in the early 1950s. Because of its depressing effects and slow action (often several days or weeks), reserpine is not presently used as a major tranquilizer, but it is the most widely used antihypertensive drug in the United States.

In place of reserpine, today, there is a large group of synthetically produced compounds called *phenothiazines*. The phenothiazines were introduced to psychiatry at about the same time as reserpine. There are now at least ten different phenothiazines under many brand names (Compazine, Thorazine, Pacatal, Sparine, etc.). These phenothiazines have been shown to be dramatically effective in the management of the mentally ill.

In general, the members of this family of drugs appear to be equally effective in combating symptoms of psychosis, including hallucinations, delusions, excitement, agitation, and aggressiveness. The use of this drug family often allows mental patients who were previously quite violent to be handled in an open ward or a general hospital or even to leave the hospital and return to society.

When used early enough, this group of compounds, even in small doses, can prevent the development of serious psychotic dysorganization (impaired and inefficient emotional organization) and hospitalization. The use of major tranquilizers appears to be required for many months or years to prevent a patient's relapse. Therefore, years after a patient appears well, he or she may be still maintained on drugs such as phenothiazine. The careful adjustment of the kinds and amount of phenothiazine drugs used results in a normally functioning individual.

The Minor Tranquilizers

Minor tranquilizers are the most widely used group of psychotropic drugs in the United States. They are used to combat anxiety, tension, and attending symptoms, such as fast heart rate, tension headaches, gastrointestinal disturbances, restlessness, insomnia, irritability, and oversensitivity. Minor tranquilizers somehow insulate an individual from external stimuli that cause anxiety and stress. It is generally believed that these drugs, which decrease excessive emotional behavior, act directly on the reticular formation and lower areas of the brain; but the exact sites of action have not been identified.

A large variety of these agents are on the market (see Frontispiece, "Drugs Commonly Abused"). Four of the most frequently used minor tranquilizers are *benzodiazepines* (Valium), *meprobamate* (Miltown and Equanil), *diphenylmethanes* (Phobex, Suavitil, and Atarax) and *chlordiazepoxide* (Librium and Librax).

The minor tranquilizers are potent antianxiety drugs and should be considered dangerous. There have been reports of both addiction to meprobamate and suicides. Thus, they are not as harmless as their advertising image and casual use by the general public might suggest.

Few of the minor tranquilizers are commonly used illegally, mainly because they do not produce any euphoric effect. But addiction to them can come from increased dosages over a long period of time. Consequently, in the addiction-prone personality the progressive increase in self-administration of tranquilizers is a very real danger and would be considered drug abuse. When large dosages are taken for long periods of time, a sudden withdrawal may result in muscular twitching, convulsion, and other withdrawal symptoms. The dosages recommended by a physician may be maintained for extremely long periods of time without adverse effects.

CANNABIS (MARIJUANA)

Cannabis (Figure 3.11) has been well known since ancient times. The drugs are produced from the many varieties of *Cannabis sativa* (Figure 3.12) grown throughout the world. The leaves (Figure 3.13) and flowering tops of the female plant secrete an amber-colored resin that contains the chemicals delta-8-tetrahydrocannabinol (delta-8-THC), and delta-9-tetrahydrocannabinol (delta-9-THC), which seem to be the active substances causing mood modifications and behavior changes in the user. *Cannabis* is probably known by more names throughout the world than any other plant. Some of the names for it (or its derivatives) are "Indian hemp," "Canadian hemp," "hemp weed," "Indian hay," loco (crazy) weed, "weed," "grass," "pot," marijuana (or marihuana), "13," "maryjane," "kif," hashish, "hash," bhang, charas, ganja, and THC.

This drug family, more than any other, cannot be accurately explained without specifying dosage levels. The potency of the intoxicating drugs produced from *Cannabis* varies widely throughout the world, depending upon the genetics of the variety being used, how it is prepared, which parts of the plant are used, and how it is stored until it is used. The marijuana used in the United States is probably the weakest preparation of the plant used in the world. The most potent is *charas* (used mainly in India), which is the unadulterated resin obtained from the female plant or its dried flowers. The term *hashish,* when used correctly, indicates a powdered and sifted form of charas (or other preparations made from charas). Hashish ("hash") and oil extracts are now being illegally imported and sold in the United States. A form of *cannabis,* delta-3-tetrahydrocannabinol (delta-3-THC or THC), has been synthetically

CANNABIS (MARIJUANA)

produced. It has been used in some scientific studies and has been found to be less potent than the naturally occurring THC. There have been reports of this material being available on the street, but it is highly unlikely. The greatest portion of THC available is of plant origin. Consequently, it is not pure and especially in the liquid state may represent an unknown broad mixture of chemicals that may be very dangerous.

In large doses (as is the case in the use of hashish or THC liquids), cannabis bears many similarities to the hallucinogenic drugs. This is why, for the last few years, it has been classified as a hallucinogen and

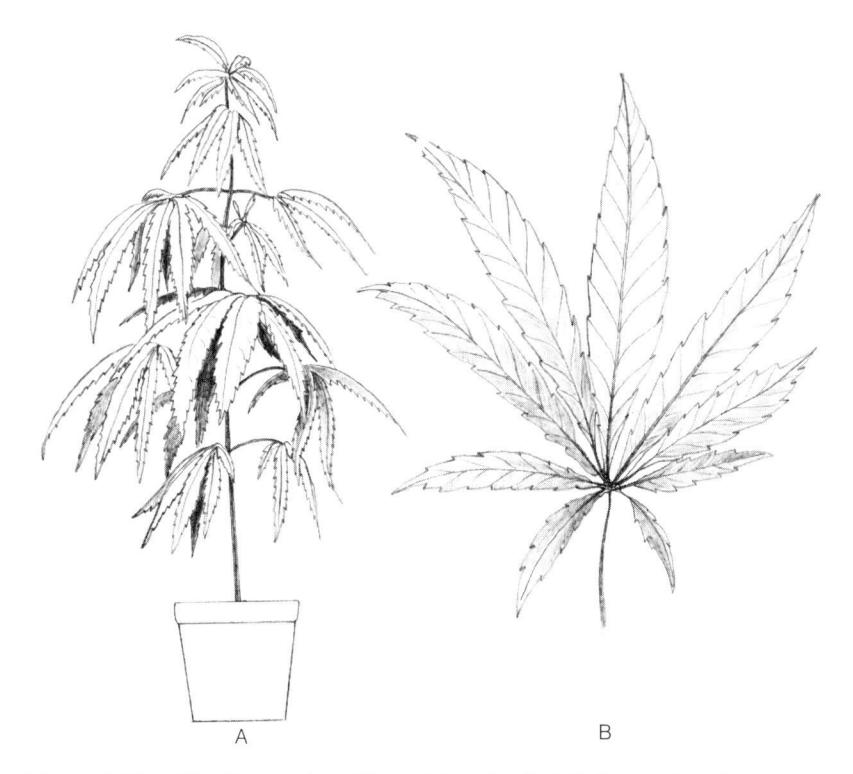

Figure 3.11. Marijuana plant (Cannabis sativa): *(A)* Cannabis sativa *drawn from a young potted plant. The live green plant has a characteristic odor, is sticky to the touch, and is covered with fine hairs that are barely visible to the naked eye. The flowers of the female plant form irregular clusters containing light, yellow-green seeds. The stalks and stems are used in the textile industry for the manufacture of rope, twine, mats, bags, and certain grades of coarse paper. (B) The leaves are compound, of from five to eleven (always an uneven number) leaflets or lobes extending diagonally from two to six inches from the center to the edges. The two outer lobes are always very small compared with the others. The leaf is deep green on the upper side and a lighter green on the lower side.*

Figure 3.12. A field of marijuana plants. Marijuana attains a height of from 3 to 16 feet.

will continue to be classified as such by many individuals. If it is so classified, it should be placed on the near end of the continuum of action section for hallucinogens, and LSD should be placed at the far end (Figure 2.9). Because of recent research and reports into the effects of *Cannabis* drugs, the authors of this book feel that it should be given a classification distinct from all other drugs. This is because of its wide range of actions, which are similar to hallucinogens (a stimulant) and to sedatives, alcohol, and narcotics (all depressants).

Effects of Cannabis

Individuals using potent extracts from *Cannabis* experience drastic distortions of auditory and visual perception, hallucinations, and a sense of depersonalization similar to that occurring with the use of LSD. At comparable doses LSD is 160 times more potent than cannabis in producing these effects. Tolerance occurs very rapidly with the hallucinogens, but there seems to be little tolerance development with cannabis. Also, there is no cross-tolerance between LSD or mescaline and cannabis.

The effects of even large dosages of cannabis are milder and more easily controlled than is the case with LSD or mescaline. The differing "trips" or "highs" of the two classes of drugs are readily distinguishable by users; cannabis users, even at high dosages, lack the major anxiety, panic, and stress reactions found in hallucinogen users.

Cannabis, at any dosage level, does not show the increased body temperature, increased blood pressure, or pupil reactions of the hallucinogens. Cannabis intoxication ends in sedation and sleep; whereas wakefulness is characteristic throughout intoxication with hallucinogenic drugs.

At low dosages, as in the use of marijuana, the effects of cannabis and alcohol are very similar. Both produce an early excitement and a later sedated phase, which appears faster in the marijuana user. The marijuana "high" is distinguishable by the user from alcoholic intoxication.

When comparisons were made of the effects of alcohol and marijuana on mood modifications and physiological functioning, it was found that they were similar in their effects except for the alteration of perceptions that was produced by marijuana but not by alcohol. Both produced decreased physical activity, drowsiness, and decreased performance on mental tests. Marijuana led to a moderate stimulation accompanied by overestimations of time and space. Comparable amounts of marijuana and alcohol produced about the same decreased performance on mental

Figure 3.13. A marijuana leaf. The leaves are compound, composed of from 5 to 11 (always an uneven number) leaflets or lobes.

tests. Also, the combination of marijuana and alcohol generally led to a poorer performance than occurred with either drug alone. Lastly, appetite and food consumption were increased by marijuana and decreased by alcohol.

The margin of safety for cannabis is far greater than that for alcohol. Large dosages of alcohol (Figure 2.10) act as a general anesthetic, producing a continuous depression of the central nervous system; whereas large dosages of marijuana tend to backtrack toward the stimulating side of the continuum.

In the United States, the most commonly used cannabinol is marijuana, the dried leaves and flowering tops of the female hemp plant. Cigarettes are made by rolling the weed in cigarette papers with the ends tucked in to prevent loss of the prickly, loosely packed weed. These marijuana cigarettes cannot be confused with tobacco cigarettes. First, marijuana is greener than tobacco. The marijuana burns hotter than cigarettes made of tobacco, and the burning tip is brighter. Also, the lighted tip goes out easily unless a continuing effort is made to keep it lit. Burning marijuana has the smell of burning hay, leaves, or weeds. Consequently, someone cannot smoke marijuana without easily distinguishing it from tobacco.

Shortly after inhaling marijuana, the user has a feeling of inner joy that is far out of proportion to his or her actual situation. This feeling is described by users as being "high." If someone smoking marijuana is alone, they may be quiet and dreamy; they may just sit and watch the passing parade of technicolor illusions that occur. In a group, they may be talkative and happy and can be easily misled regarding abilities and intellectual capabilities. As we have explained, senses of touch and perception are changed; and the ideas about time, space, and speed are distorted. The user's coordination is altered, although they, or others, may fail to recognize either this alteration or the impairment of intellectual capacities. All behavioral modifications depend upon environment, personal feelings, and the amount of marijuana smoked. If a user is in a negative mood or in unpleasant surroundings, they may become anxious and apprehensive. If they continue smoking in this environment, they may become easily irritated. With increased use at this point, the smoker may become confused, disoriented, and afraid. Behavior can become impulsive and mood reactions highly variable and unpredictable. With increasing dosages (Figure 2.10), they may at this point experience hallucinations and other reactions associated with the more potent hallucinogens. The total effects of a marijuana experience last from 3 to 5 hours, after which the user feels slightly drowsy or sleepy and hungry.

Little research has been completed that could help explain the actions of cannabis within the central nervous system. Various animal experiments have produced the following results: Cannabis can reduce acetylcholine action, which may account for the decrease in nerve impulses in integrative centers (Figure 2.8) of the brain. It potentiates norepinephrine, and this may be the action that results in its potentiation of amphetamines when the two are taken together. (See the section on amphetamines later in this chapter for a description of how this may take place.) But it does not increase the amounts or activities of norepinephrine, which is paradoxical because it does decrease serotonin amounts (Figure 2.8), and an imbalance between these two usually causes greater norepinephrine stimulation, thus producing a greater increase in sympathetic nervous activity. There is an increase in some areas under sympathetic control, but this does not constitute uniform stimulation throughout the system. The action of cannabis seems to be very specific and transient. It has been shown to stimulate frontal areas of the cortex while depressing parietal (side) areas; but one hour after administration, both areas become depressed. This action could account for what has been termed "temporal disintegration" (difficulty in retaining, coordinating, and indexing serially in time those memories, perceptions, and expectations that are relevant to the goal being pursued). Marijuana smokers experience difficulties in speech, are unable to remember from moment to moment the logical thread of a conversation, cannot retain events from the preceding few seconds or minutes, cannot shift attention appropriately from one focus to another, and cannot organize and coordinate a time sequence of recent information.

The ability to arouse the cortex is diminished, which suggests further actions within the integrative centers of the hypothalamus, the limbic system, and the reticular activating system. High accumulations of cannabis compounds have been recovered from the areas of the limbic system of hallucinating animals. This is also the case with animals experiencing behavioral and mood modifications. Marked amounts of cannabis have been found in the cerebellum of animals during periods of muscular-coordination dysfunction.

Thus, it appears that the actions of cannabis in the brain are widespread and related to increasing and decreasing concentrations of cannabis and its metabolic by-products in specific brain areas. The specific actions are still unknown, and no specific avenues of actions are yet apparent.

Controlled research has just begun to expand the available knowledge concerning marijuana. Since the National Institute of Mental Health embarked on a high-priority cannabis research program, an adequate supply of standardized natural and synthetic substances has been developed. As time passes, additional information will become available to provide a more complete picture of the implications of marijuana use at various dosage levels and in differing patterns.

HALLUCINOGENS

A drug that creates vivid distortions in the senses without greatly disturbing the user's consciousness is called a *hallucinogenic* drug. These distortions vary greatly but frequently include hallucinations. In the popular literature, a number of other terms have been used to describe these drugs and their effects, such as psychotomimetics, psychedelics, or pseudohallucinogenics (producing false hallucinations). The terms *psychotomimetic* (psychosismimicking) and *psychotogenic* (psychosis-producing) have been used in reference to some hallucinogens because many individuals under the influence of these drugs exhibit behavior that resembles the disturbances seen in the psychotic or severely mentally ill. Some of the drugs seem capable of temporarily producing psychotic behavior in a normal person. It has been necessary to hospitalize some users to prevent them from doing harm to themselves or others during their temporary psychosis.

A large number of individuals have acquired these drugs illegally and have taken them without medical supervision in cult-like group experiences, in small groups (two or three individuals), or alone.

The hallucinogenic drugs are placed only slightly above the neutral area on the continuum of drug effects (Figure 2.9) because in what would be considered therapeutic or minimal dosages, all members of the group produce rather mild stimulation of the individual. Within the hallucinogenic group, different classes of drugs produce a wide continuum of stimulation and hallucinogenic effects. Many of these drugs, such as LSD, produce extreme reactions in almost incomprehensibly small dosages.

An explanation of the actions of hallucinogenic drugs is still rather fragmentary. The chemical structures of LSD and serotonin are very similar. Consequently, authorities feel that the action of the hallucinogens is tied to serotonin. One theory is that it either blocks the production of serotonin, upsetting the balance between it, acetylcholine and norepinephrine, or replaces serotonin, flooding the brain stem with a serotonin-like substance. There are electrophysiological changes during and after the use of hallucinogens that can be recorded from the cortex to the stem of the brain. These show increased activity of the central nervous system toward an alert or arousal pattern of behavior. There also seems to be an increase in excitability (which may actually be suppression of opposing inhibitor systems). And there is a drastic change in ability to receive sensory stimuli. Some researchers have described this as a continuum of brain excitability from arousal to excitatory blocking of senses and sensations to disorganization of brain function.

It has been suggested that the combination of brain arousal and impaired sensory reactions leads to the overawareness of previously stored information processed by the brain (called a "preconscious" stream of information) which the individual is then made aware of in experiences ranging from fragmentary images to well-developed, extremely complete epics. An example of these actions could be the suggested effects of LSD. It increases and scrambles activity at sensory synapses (causing the changes in ability to receive sensory stimuli), is a general arouser of the reticular formation and associated areas, and seems to be able to release vast preconscious streams of information.

There are two major chemical classes of hallucinogenic drugs: the phenylethylamines (which are very similar to norepinephrine) and the indole alkaloids (which are very similar to serotonin). Tolerance to both the groups develops very quickly, and cross-tolerance within the groups also develops.

Phenylethylamine Drug Family (Mescaline)

The most significant drug abused in this group is *mescaline* (trimethoxyphenethylamine), named after the Mescalero Apache Indians of the southwestern United States, who developed the cult of peyotism. The drug mescaline occurs in peyote, which is a small, spineless cactus that grows naturally in northeastern Mexico and the watershed of the Rio Grande. Peyote is carrot-shaped, with only the topmost part extending above the ground (Figure 3.14). This portion is cut off, and although it may be eaten fresh, it is usually dried to form the peyote, or "mescal," buttons. It may also be boiled and the broth drunk.

Today, peyote buttons are chewed mainly by Indians of a number of tribes in the southwestern United States to induce hallucinatory states used in their religious rituals. In the ceremonies of the Native American Church, individuals consume peyote ritualistically, beginning about an hour before important religious festivals. This legal use of peyote makes the control of its distribution difficult, since the restrictions of its use in this situation has been decreed by the U.S. Supreme Court as "a restriction of an individual's freedom of religion."

Mescaline is of interest because it is chemically related to the amphetamine compounds and to norepinephrine, a natural substance in the body. This chemical association seems to be responsible for the effects mescaline produces on the autonomic nervous system prior to the onset of psychic reactions. Soon after taking mescaline, any psychic experience is preceded by 1–3 hours of flushing, vomiting (which greatly limits the abuse of this drug by young people), cramps, sweating, increased pulse rates, elevated blood pressure, muscle twitching, and other autonomic phenomena. These effects are followed by 4–12 hours, or several days in some cases, of visual hallucinations in all ranges of color and fantastic geometric patterns, vast feelings of depersonalization, and great distortions in the sensing of time and space relationships.

Mescaline is usually ingested in the form of a soluble crystalline

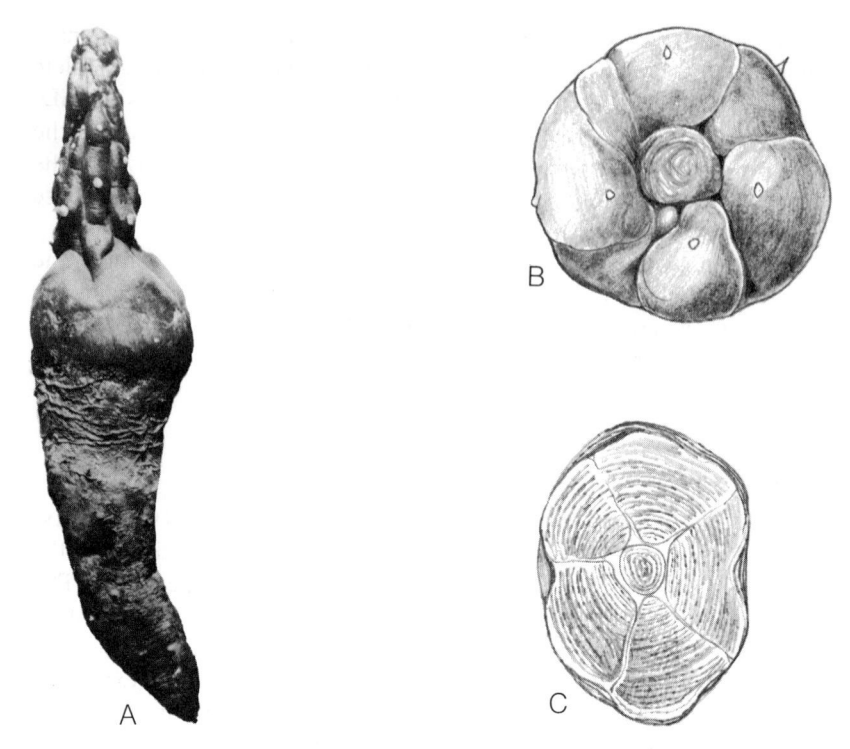

Figure 3.14. Peyote cactus plant (Lophophora williamsii): *(A) Complete plant. (B) Top view of rounded top of a plant, called a peyote or mescal "button." (C) Bottom view of a severed "button."*

powder, which is dissolved, or in a capsule. It must be used in high dosages in order to produce the extreme hallucinogenic effects usually sought. These dosages may be dangerous to many individuals who have had no warnings of the impending psychic reactions.

Of the other drugs in this group (anhalonine, TMA, MDA, MMDA, etc.) only one, DOM or "STP," has been used to any extent. "STP" is the street name for a drug identical to DOM, an experimental drug produced by Dow Chemical Company. It is known to have been used in San Francisco during the summer of 1967 for a short period of time. Whether "STP" was a theft of Dow's formula or supply or was independently produced was never determined. Ten cases with "bad trips" were admitted to hospitals in 11 days; one chronic psychosis was produced, and one death may have been caused by the drug. Some feel it carries a high risk of psychotic reaction, hence the saying "STP kills."

Indole Alkaloids

There are a number of drug families (ergot alkaloids, tryptamines, and the *harmala* group) and individual drugs (DMT, bufotenine, morning-glory seeds, psilocybin, LSD, and others) that make up this group. Many are found naturally occurring in plants and some in animals. Also, indole alkaloidlike compounds with hallucinogenic potentialities have been synthesized in the laboratory or produced by rearranging the chemical structure of the naturally occurring substances. This rearrangement and synthesis may have almost limitless possibilities, which, in the future, could deluge humanity with extremely potent and highly selective drugs affecting the central nervous system in as-yet-unknown ways.

DMT (dimethyltryptamine) is found together with *bufotenine* (which was originally isolated from the skin of toads) in the Caribbean cohoba bean, chewed by certain Indians in South America to produce religious illusions and visions, and in the seeds of the domestic morning-glory plant, abused in the United States for their hallucinogenic effects. *Psilocybin* is the active ingredient of the ritually employed hallucinogenic mushroom *Psilocybe mexicana* of central Mexico. Another drug, *harmaline,* is isolated from shrubs and used by South American Indians to produce hallucinatory states. Also, a similar compound, *ibogaine,* is used by African natives to remain motionless for as long as two days while stalking game. Hallucinogenic compounds are universal in their distribution and have been known for many years.

Probably the most potent indole alkaloids are the *ergot alkaloids*. These were first isolated from a fungus found on rye and some other grasses (Figure 3.15). It is believed that these alkaloids in the fungus were probably responsible for the convulsions, mental confusion, and gangrenous changes in the lower limbs associated with periodic outbreaks in the Middle Ages of Saint Anthony's Fire, which was probably produced by eating foods containing infected rye.

All ergot alkaloids can be changed into *lysergic acid*. Various derivatives of this compound have been developed. But the most potent and the most famous is *D-lysergic acid diethylamide-25* or *LSD 25*. LSD is such a potent drug (over 800 times more potent, at a similar dosage level, than mescaline) and is abused in such small dosages (as low as 1 microgram [0.000001 gram] per kilogram [2.2 pounds] of body weight in man) that it is almost impossible to produce anything but an extreme reaction (Figure 2.10 and 2.11) when taking illegally prepared dosages. The average abusive dose is between 100 and 250 micrograms. Since LSD ("acid") is the best known, we will use it as the example in explaining this group of hallucinogens. LSD is a tasteless, colorless, and odorless drug. The physical side effects and complications seem to be discomfort: nausea, vomiting, aches, and pains. It dilates the pupils, raises the blood pressure, and increases the strength of the reflexes (by actions on synapses within the central nervous system). It stimulates cerebral sensory centers and blocks off inhibiting mechanisms within the reticular formation and associated areas. By its actions upon sensory centers, LSD produces visual hallucinations in all ranges of color. Actions

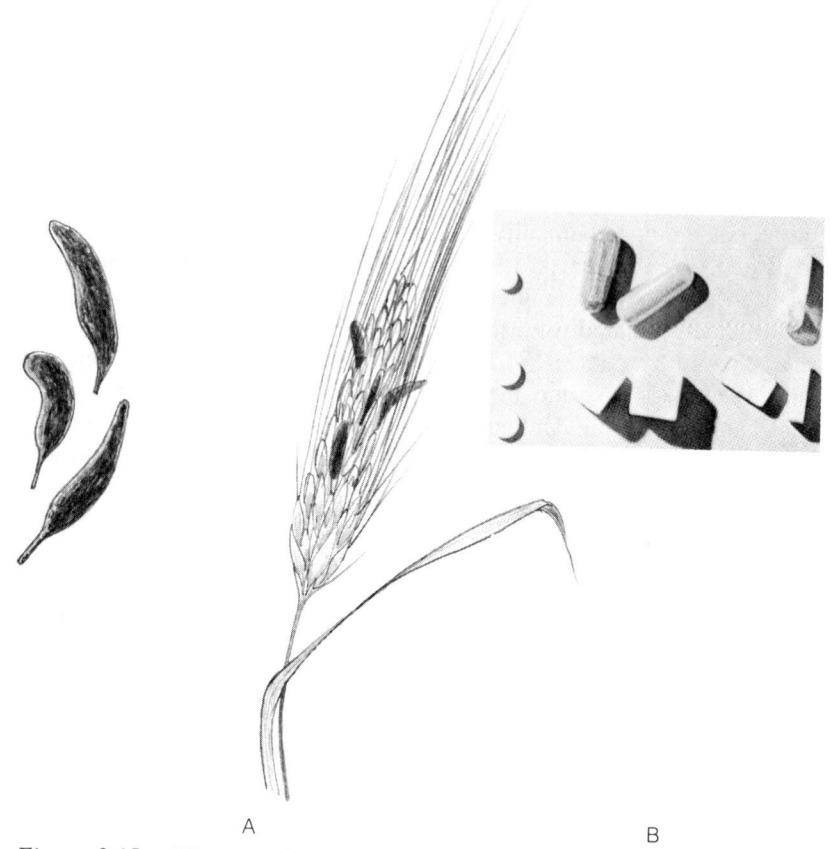

Figure 3.15. The ergot fungus Claviceps purpurea *and illicit forms of LSD: (A) The ergot fungus as it might be located on the head, or spike, of a rye plant. The drug LSD (lysergic acid diethylamide) is derived from this fungus. (B) Illicit pills and capsules of LSD.*

within these areas of the brain intensify hearing, increase the sensitivity for feeling textures, and may produce a tingling sensation and a numbness of the hands and feet. Increases in the ability to taste and smell are not frequent. Subjects often report crossover (scrambling at sensory synapses) of sensation; for example, they may "hear" colors or smell the "scent" of music. There are also reports of individuals being badly burned because flames felt cool to the touch.

Following the initial physical effects of the drug, users show a great degree of subjective euphoria and feel a sense of mental clarity or comprehension, although objectively they appear to an observer to be confused, uncoordinated, hallucinating, and disoriented. Upon recovery from a profoundly moving and significant experience, they often have a personal (subjective) feeling of being reborn that may be accompanied by a sense of deep affection for others who were present and participating in the experience.

LSD was first used experimentally to produce an artificial psychosis (termed a "model psychosis,") resembling an acute schizophrenic reaction and lasting for several hours. The temporary modifications within the nervous system caused by LSD may produce any type of behavioral disturbance. Anxiety and panic may occur during the user's struggles to maintain control of the situation, and fear reactions can occur because of the user's distorted time and space perceptions. Suicidal attempts are common in those who experience panic or anxiety during LSD abuse (called a "bummer" or a "bad trip"). An overdose (toxic dose) of LSD may produce long periods of delirium, convulsions, or "flashbacks," in which the psychoticlike effects of the drug keep coming back at intervals for up to a year or more. These prolonged reactions to any hallucinogen (mescaline, LSD, and similar substances) seem to be caused by a mixture of the drug's effects and the psychiatric disorders that were already present in the users prior to their use of drugs. However, no one has been able to predict who is likely to have such an extreme reaction or when they will have them.

Because of the large number of extremely disturbed reactions that are occurring with the illegal use of LSD, it is possible that much of the illegal supplies of LSD and possibly other extracted or synthetically produced hallucinogens are either contaminated or mixed with other substances. Since the legal production and distribution of LSD and similar compounds is closely controlled, very little, if any, is funneling into the illegal market. The illegal supplies are being produced in foreign countries and in basement laboratories here in the United States.

Both the single experiment and the chronic use of LSD present a danger to mental health. The LSD experience is so variable and so complex that a single explanation of its use will not always hold true. Discussions conducted by Dr. Sidney Cohen, chief of psychiatric services at the Wadsworth Veterans Administration Hospital, in a meeting in San Francisco in 1966, explained in detail the many different reasons for trying LSD and described personality types who have used LSD. The motives of LSD users range widely. These include an adventurous desire to have new experiences or the thrill of a shared, forbidden activity with a group to provide a sense of belonging. Many are searching for sensual experiences. Some are making genuine attempts to achieve greater self-understanding and self-fulfillment and have a need for a truly mystical experience. This is especially true of persons with a philosophical conviction inclined toward the transcendental (seeking of an ultimate truth beyond natural or rational comprehension). Also, there is a search for the fulfillment of a private utopian myth or, with many, a desire to attain greater creativity. But as Dr. Cohen has been heard to say many

Figure 3.16. Dried coca leaves and illicit forms of cocaine commonly available.

times. "Creativity is ninety percent perspiration and ten percent inspiration, and LSD does not enhance one's desire to sweat."

COCAINE

Cocaine is the principal active ingredient of the South American coca plant (*Erythroxylon coca*) shown in Figure 3.16. The coca plant has been cultivated since prehistoric times in the Andean highlands of Bolivia, Chile, Colombia, Ecuador and Peru. In these areas its leaves are chewed for refreshment and relief of fatigue, much as North Americans once chewed tobacco. At these dosages the user experiences relief from fatigue and increased capacity for work, feelings of physical prowess, and a sense of optimism about abilities and achievements.

Small amounts of coca leaves are grown and legally exported by Peru and Bolivia to the United States and Europe, where the leaves are "decocainized" and then used as flavoring extracts in soft drinks. The extracted cocaine itself is sometimes used as a local anesthetic in operations of the ear, eye, nose, and throat. Other uses of cocaine (such as a dental anesthetic) have been supplanted by synthetic drugs such as Provaine, Novocaine, or Xylocaine.

Processed cocaine is an odorless, white, fluffy, fine crystalline powder, similar to snow in appearance. Consequently, on the criminal market, it is commonly referred to as "snow" or "coke." The majority of illicit cocaine is extracted from coca leaves in the Andean highlands and converted into coca paste. This paste is then refined into the powdered crystals in Bolivia and Peru, from where it is smuggled into the United States.

As shown on the *Continuum of Drug Effects and Actions* (Figure 2.9) this drug is the most powerful natural stimulant known. In recreational dosages it affects the higher centers of the brain; then, as it is used more frequently or as dosages are increased, its stimulating effects stimulate downward (cerebral cortex, brain stem, then medulla).

Stimulating effects on the cerebral cortex result in euphoria, laughter, restlessness, and excitement. During this phase an individual is talkative, active, and excited. As dosages are increased, the lower centers (thalamus, hypothalamus, and reticular formation) are stimulated, pulse increases, blood pressure elevates, respiration speeds up and deepens, and the arousal and wakefulness reactions are stimulated. Continuing to take the drug at closer intervals or increasing the amount being used at one time causes hallucinations, confusion, loss of muscular coordination, tremors, and convulsive movements. A quick depression ("letdown") takes place following the period of stimulation. The higher centers are depressed first, and euphoria is replaced by anxiety, headache, dizziness, and fainting. The individual may then plunge into stupor, sleep, or coma. These signs are characteristic of an overdose, and death due to respiratory failure may follow because of the effect of the cocaine upon the respiratory center in the brain.

There are two distinct types of cocaine poisoning. In the first, death is due to *circulatory collapse*. This may occur in any person after relatively small doses and is due to a *drug idiosyncrasy* (an abnormal response of a person to a drug). There is no way to tell when or where this will take place. Such circulatory collapse is recognized by pallor, dizziness, nausea, failure of the pulse, and loss of consciousness. The second type of cocaine poisoning is the overdose. It is characterized by delirium, increased reflexes, convulsions, and violent manic behavior. Respiration is at first stimulated and then depressed. Death is due to respiratory failure. Cocaine also stimulates the temperature-controlling center and the vomiting center in the brain. This results in sweating and vomiting. When sweating can no longer occur because of the constriction of blood vessels in the skin, there may be a dangerous elevation in body temperature which may also produce death.

Cocaine is usually taken by "sniffing" [called "snorting"] it into the nostrils or by injection. It is rarely swallowed because it loses its potency when taken by mouth. It produces almost violent stimulation and euphoria when injected. Consequently, sniffing is the most popular method because the drug is absorbed through the membranes of the nose slowly, prolonging the stimulating effects and tempering the drug's intensity and drastic effects. The euphoria, or "rush," is relatively short-lived, and compulsive users may inject or "snort" small dosages as often as every ten minutes. Significant tolerance develops with cocaine, and when a habitual user stops, he or she drops into a deep depression very similar to the amphetamine "crash."

Other drugs belonging to the same family as cocaine (piperidine derivatives) have been chemically rearranged in the laboratory into very potent drugs. The complete group of drugs has some of the effects of hallucinogens and some of the effects of amphetamines. These drugs are also similar chemically to amphetamines (some authorities classify cocaine as an amphetamine). At times, cocaine is also classified as a hallucinogenic drug. The authors have placed it slightly above the hallucinogens and below the amphetamines because of the similarities of its effects.

ANTIDEPRESSANTS

Psychic energizers, antidepressants, and amphetamines can all be used to combat depression. A whole series of names (including analeptics, thymoleptics, and depressolytics) have been invented to describe them, but the three names given here are more useful. These three names correspond to relative actions and chemical structures within the group. However, all three types of drugs could be labeled with the simplest term: antidepressants.

The actions of these drug groups on the central nervous system are based on what is known as the *amine theory of mood*. This theory postulates that the amount of availability of three hormones in the brain is related to the mood of an individual. The three substances are serotonin, in this instance acting as a synaptic transmitter in some areas of the central nervous system, norepinephrine, an activator of the sympathetic nervous system, and dopamine, a generalized nervous system activator that is changed by the body into norepinephrine. These three substances are known chemically as *monoamines* because they contain a single amino group in their chemical structure (Figure 3.17). The theory contends that the higher the brain levels (or availability for neural transmission) of compounds containing an amine, such as these, the happier, more alert, and more energetic a person will be. The lower the amine levels, the more depressed and apathetic the person is.

The actions of many of the previously discussed drug groups seem to

ANTIDEPRESSANTS

fit this theory. For example, the major tranquilizers, such as reserpine and the phenothiazines, tend to deplete the brain of amines and are associated with the production of depression. Also, hallucinogens produce high amine levels within the central nervous system. And the ability of phenothiazine compounds (such as Thorazine) to reverse hallucinogenic experiences is probably due to their ability to deplete these amine levels in the central nervous system.

Psychic Energizers

Psychic-energizing compounds are mild cerebral stimulants and produce an elevation of mood, increased activity, heightened confidence, and an increased ability to concentrate. They are used in treating depressed states that are not associated with compulsive behavior or pronounced anxiety, for example, in elderly individuals who feel that life no longer

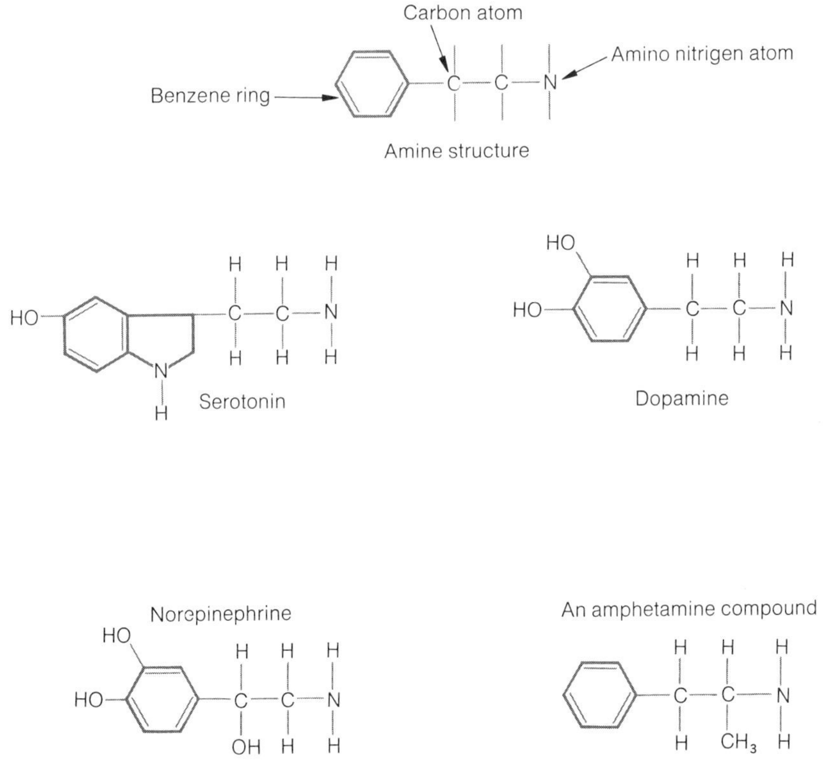

Figure 3.17. Chemical structures of important natural body "amines." "Amine" compounds all have a similar chemical structure—a benzene ring and an amino nitrogen atom separated by a chain of carbon atoms.

holds anything for them and in persons ill with chronic diseases (such as terminal heart disease, cancer, and emphysema).

The psychic energizers counteract drowsiness, inactivity, indifference to surroundings, intense worry, general depression, and oversedation associated with the side effects of major tranquilizers, barbiturates, and some antihistamines.

As previously noted, according to the amine theory of mood, the mood of an individual is thought to be greatly influenced by the relative levels of three hormones (serotonin, norepinephrine, and dopamine) within the nervous system. This is physiologically controlled by an enzyme, *monoamine oxidase,* that inactivates and destroys these neurotransmitters within the body. Thus, the enzyme is able to control the amounts of these transmitters and indirectly the mood of an individual within normal or abnormal limits.

The major psychic energizers are termed *monoamine oxidase inhibitors* (MAO inhibitors). These drugs (Marplan, Niamid, Nordil, and Parnate) bring about an increase in serotonin and norepinephrine in the central nervous system by inhibiting the destroying and inactivating actions of monoamine oxidase. It is speculated that the MAO inhibitors irreversibly inhibit the action of this enzyme, thus promoting the accumulation of the neural transmitters, which in turn stimulate the mentally depressed individual and elevate mood.

These drugs are very dangerous and should be administered only under constant medical supervision. They potentiate almost all other mood-modifying drugs, and deaths have resulted when they were taken with alcohol, narcotics, anesthetics, and amphetamines. Also, deaths have been caused by eating cheese [see "Tyramine—rich foods" of Table 1.1] while taking MAO inhibitors.

Antidepressants

The major differences between the antidepressants and the psychic energizers are the degree of stimulation, their chemical actions, and the degree of side effects produced. Both are often classed together as antidepressants. The psychic energizers relieve the outward signs of depression effectively, but the inner feelings of the individual remain less changed. Consequently, the person appears active and changed, but the unhappy psychology remains. Frequently, such individuals try to commit suicide because they have recovered sufficient physical energy to carry out suicidal thoughts and actions. The antidepressants tend to make a person become more cheerful, active, and interested and also to relieve hopelessness, despair, fatigue, and guilt to a greater degree; but the chance of suicide still remains.

Antidepressants (such as Tofranil and Elavil) act by blocking a major route of inactivation by monoamines (reuptake by certain brain cells), allowing nerve cells to react for prolonged periods of time and thus increasing the stimulating effects.

Anticholinergics

A group of stimulants termed *anticholinergics* produce actions similar to those of the antidepressants. Two (atropine and scopolamine) are extracts from natural sources, and a few others (including Mesopin, Pamine, and Banthine) are synthetics. These drugs block or inhibit the actions of acetylcholine (Figure 2.8) and exert a stimulating effect of their own on the parasympathetic division of the autonomic nervous system, relaxing the muscles of the body and decreasing anxiety and stress. In abusive dosages, they are quite dangerous, causing hallucinations, drastic personality distortions, paralysis of the muscles of the body, and in a number of instances, death. Scopolamine is found in some over-the-counter sleeping agents. These have been abused by some individuals. Also, drugs within this group have been sold to individuals as something else (hallucinogens such as LSD or amphetamines).

Atropine and scopolamine are extracted mainly from *Atropa belladonna,* deadly nightshade (Figure 3.18), and are sometimes called "belladonna." They stimulate from the cortex down (cortex to reticular activating system to medulla). They cause an individual to become restless, wakeful, and talkative. If the dosage is high, or if someone continues to take doses, the stimulation can continue into delirium, coma, and death (Figure 2.9). Atropine stimulates respiration at first, then depresses it until death from respiratory failure takes place. Scopolamine and some of the synthetics cause more euphoria, relief of fear, and relaxation than does atropine; but some cases of amnesia have been caused by these drugs.

AMPHETAMINES

The amphetamines (Figure 3.19 and Frontispiece, "Drugs Commonly Abused") are a large group of synthetic drugs that stimulate the central nervous system, particularly the reticular formation and the cerebral cortex. In addition, they exert effects similar to those obtained when the sympathetic nervous system is stimulated. This is because the primary action of amphetamines is increasing the release and blocking the deactivation of *catecholamines* (dopamine, norepinephrine, and epinephrine). Catecholamines are the neurotransmitter substances which activate the *ergotropic division* within the hypothalamus and the *sympathetic nervous system* (Figure 2.8). Drugs that activate the sympathetic nervous system are called *adrenergic drugs*; an older term, still used at times, is *sympathomimetic drugs.*

Chemically, amphetamines release catecholmines from storage sites

Figure 3.18. "Deadly nightshade" (Atropa belladonna). *The common name given this plant explains its poisonous qualities.*

in the body. Thus, they are thought to produce their stimulating effects, in part, by flooding brain synaptic sites with such neurotransmitter substances conforming to the amino theory of mood modification. This would explain their very potent psychic-energizing, antidepressant, and sympathomimetic effects. Also, the postamphetamine depression (Figure 3.20), common in the use and abuse of amphetamines, has been explained by the norepinephrine-deficient state in which the brain is left following the abuse of amphetamines.

In general, the physical reactions amphetamines produce are signs of sympathetic nervous system stimulation, for example, an increase in heart rate, a constriction of certain blood vessels, an increase in blood pressure, a dilation of the pupils, an increase in the breathing rate, an increase in sweating, and a cottonlike dryness in the mouth. These autonomic reactions are always combined with the primary action of amphetamines on the brain. The user has both an increase in bodily activity and an arousal and elevation of mood. The arousal and elevation of

AMPHETAMINES

mood is often one of increased confidence, euphoria, fearlessness, talkativeness, impulsive behavior, loss of appetite, and a decrease in fatigue.

There are on the market a large number of amphetamine drugs that are commonly used for weight reduction (diet pills), such as the widely abused amphetamines Benzedrine and Dexedrine. On the illegal market they are known as "whites," "bennies," "dexies," "uppers," or "pep pills" (Figure 3.19). Several drug companies, without showing substantial evidence, make claims that their particular compound suppresses the appetite without causing central nervous system stimulation. No amphetamine or amphetaminelike compound has only one of these two actions in the body. Consequently, a common circumstance is that although the user loses weight; he or she also loses sleep. At first, these amphetamines are useful to depress the appetite of someone who is dieting to overcome obesity, but tolerance to their ability to block the

COLOR AND SHAPE OF CAPSULE OR TABLET	TRADE NAME	STREET NAME
Red-pink	Benzedrine (spansule capsule)	"Bennies"
Pink	Benzedrine (tablet)	"Bennies"
Orange	Dexedrine (spansule capsule)	"Dexies"
Orange	Dexedrine (tablet)	"Dexies"
Green	Dexamyl (tablet) (contains dexedrine and amobarbital)	
White	Edrisal (tablet) (contains benzedrine, aspirin, and Phenacetin)	
White	Biphetamine (capsule)	"Whites"
White	Methedrine (tablet)	"Meth" "Speed" "Crystals" "Whites"

Figure 3.19. Commonly used and abused amphetamines. The primary effects of the many brands of amphetamines—available in white or colored tablets or timed disintegration capsules—include an increase in confidence, euphoria, feelings of fearlessness, talkativeness, impulsive behavior, loss of appetite, and decrease in fatigue. Variously called "leapers," "uppers," "beans," "pep pills," and "diet pills," they produce strong psychological dependence.

DRUGS COMMONLY ABUSED

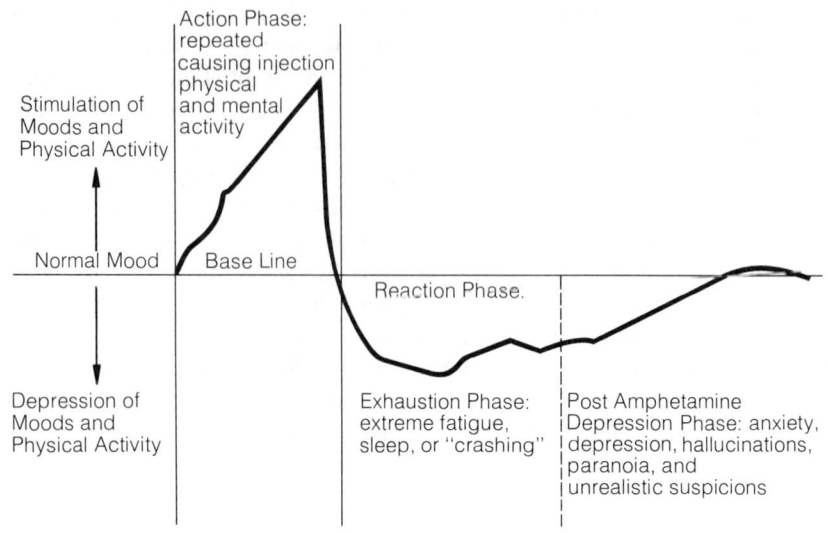

Figure 3.20. The speed cycle. (Adapted from David E. Smith, "The Characteristics of Dependence in High-dose Methamphetamine Abuse," International Journal of the Addictions, 4 (September, 1969): 453–459.)

appetite develops very quickly, making them useless. Because of this, many physicians do not recommend their use for this purpose; as soon as the drug become ineffective, the former appetite returns, and unless the person has established a new diet pattern, he or she usually returns to the old habit of overeating.

Amphetamines produce weight loss mainly by making people more active. If they keep taking more, they can keep going for hours or even days without sleep or rest. Consequently, these drugs are abused by people who want to work or play harder or longer than their normal capacities allow. To reduce this abuse of amphetamines, in 1970 the Food and Drug Administration (Drug Enforcement Administration) limited the legal use of amphetamines to three types of conditions: narcolepsy, hyperkinetic behavior (as observed in hyperactive children), and short-term weight-reducing programs. Currently, laws are being prepared which may completely remove amphetamines from the market.

The major amphetamine compounds used in this limited legal use has been Ritalin (methylphenidate) for the treatment of both narcolepsy and hyperkinetics, and Preludin for the short-term treatment of weight-loss. These three drugs are the most widely abused of the amphetamine family.

Narcolepsy is a very rare disorder in which an individual has sudden, uncontrollable desires for sleep, often as many as a hundred times a day. Amphetamines block these patterns and keep the individual awake.

Hyperkinetic or *hyperactive* children have an unusually short attention span, are unable to sit still, and in spite of normal or superior intelligence are frequently underachievers in school. Amphetamines have the paradoxical effect in such children of acting as a tranquilizer, increasing attention span, and decreasing hyperactive behavior. Considerable controversy has been focused on the treatment of hyperactivity with Ritalin and other amphetamines. *Caffeine* (in coffee) has been found to be as effective as other drugs in treating some hyperkinesis. Also, the symptoms of some hyperactive children have been relieved by keeping them from eating foods that contain artificial sweeteners and flavorings.

Many individuals occasionally take small dosages of amphetamines orally to reduce fatigue, elevate mood, produce prolonged wakefulness while doing an unpleasant task, help recover from an alcoholic hangover, or just to "get high." Others use them together with barbiturates or alcohol for "kicks." This multidrug use is physically dangerous; it can cause death or lead to impulsive acts of poor judgment and to accidents.

These drugs can cause physical damage when they are abused over long periods of time. Prolonged use of increasing dosages can cause long periods of sleeplessness and brain damage resulting in mood and behavior changes that may develop into a severe mental disorder (psychosis). People suffering from such disorder are often characterized by extreme activity for long periods of time, feelings of superiority (grandiosity), bizarre forms of suspiciousness, hallucinations, and excitement, all to an exaggerated degree.

Another widely abused amphetamine compound is *Methedrine* or *Desoxyn* (methamphetamine hydrochloride), commonly called "speed," "meth," or "crystals." Some swallow "meth" pills or inject the compound into a muscle or vein to get a quick euphoric "flash" or "rush." With continued injections, the individual is thrown into the *Speed Cycle,* shown in Figure 3.20. David E. Smith, medical director of the Haight-Ashbury Free Medical Clinic in San Francisco, California, describes the "speed cycle" as an "action–reaction" sequence of phases which take place during continued use of amphetamines.

During the *action phase,* users stay awake for days, eat very little, and experience extreme physical and neurological stimulation until their bodies become completely exhausted. While this is taking place (shown by the ascending curved line in the cycle), the catecholamine stored in the body is completely depleted. When this has taken place, no matter how much amphetamine is used, the effects the person enjoys will not be produced. Then the worst part of the speed cycle begins—the *reaction phase.* As the body adjusts to the greatly reduced amounts of catecholamines, the *exhaustion* phase sets in. Withdrawal from the drug begins; or "crashing" takes place. Users stop their injections and slip between coma and sleep for days. As catecholamine storage is reestablished, the user awakens into the second reaction phase—depressive symptoms or *postamphetamine depression*. This is a period of prolonged lethargy, deep psychological depression, nightmares, episodes of restlessness, and often long, exhausted sleep. The individual exhibits five adverse psychological reactions:

1. Exhaustion syndrome—intense feelings of fatigue. The individual may sleep continuously for one or two days.
2. Anxiety reactions. The individual becomes fearful and has unrealistic concerns about his or her physical well-being (hypochondria).
3. Amphetamine psychosis. The individual misinterprets the actions of others, hallucinates, and becomes unrealistically suspicious (paranoia).
4. Prolonged depression.
5. Prolonged hallucinations. The individual continues to hallucinate long after the drug has been completely metabolized.

During this time the storage of catecholamines is reestablished within the body and tolerance diminishes; then, when tolerance to the drug has disappeared, the abuser awakens and starts his or her injections again.

Amphetamines are frequently classed as addictive drugs, even though they do not produce a classic withdrawal similar to that produced by depressant drugs. But they do produce a strong tolerance in the individual and a strong drug dependence, and the withdrawal seems to be caused by the massive accumulated fatigue built up in the individual and the lack of norepinephrine storage for normal nerve function. Those who stop using amphetamines suddenly (often because they are no longer effective) usually go through the *exhaustion phase,* or if they are very heavy users, the *post-amphetamine depression phase* of the *Speed Cycle.*

POLY DRUG USE

Instead of a single drug being used as a recreational drug, it is common to find that many substances are being taken simultaneously or in sequence. Poly drug users take anything—known or unknown—and often start with either alcohol or marijuana. Most often, different depressants are taken together, or a depressant is taken with a stimulant.

Combinations of depressants taken together are particularly dangerous because they *potentiate* each other's actions, causing a *synergistic* effect. Sometimes barbiturates are "dropped" with alcohol or with amphetamines such as Benzedrine, Dexedrine, or Preludin. These combinations overcome the depressing effects of the barbiturates or alcohol and increase their antagonistic actions. This use of a stimulant drug to antagonize a depressant drug is dangerous because such combinations interfere with the body's normal disposal of both alcohol and barbiturates through the liver, causing a toxic or lethal level of each to be reached very quickly. Consequently, the consumption of even small amounts of barbiturates or alcohol and stimulants is dangerous, and combinations of barbiturates or narcotics and alcohol can produce extreme intoxication, stupor, and even death.

SUMMARY

I. Most people take drugs for medical reasons.
   A. Misuse of these drugs is usually a mistake or accident.
   B. Drugs voluntarily abused are those that can modify the moods and behavior of an individual.
      1. Some call these "recreational" drugs.
      2. Such drugs either depress the nervous system or stimulate it.
   C. This chapter will proceed along the continuum of drug actions and describe each drug group from the more potent depressants to the more potent stimulants.

II. Anesthetics
   A. These are the most potent of nervous system depressants.
   B. General anesthetics
      1. Early in this century *ether* and *chloroform* were used as recreational drugs. Many died in the process.
      2. Currently the most widely abused anesthetic is *phencyclidine* (PCP).
         a. It is sold as white powder and is called "angel dust," or in pills or capsules as "peace pills."
         b. It is not prescribed for humans because the range between an effective dosage and an overdose is too narrow for the general safety of the patient.
         c. Persons taking PCP hallucinate and feel "spaced out." They become aggressive, uncoordinated, and sometimes unable to speak.

III. Narcotics
   A. The noun *narcotic* comes from a Greek word meaning "to benumb."
      1. They are defined as "having the power to produce sleep or drowsiness and to relieve pain."
      2. They consist mainly of opiates and their synthetics.
   B. Effects of narcotics
      1. They are used medically for their ability to produce analgesia and sedation.
      2. All narcotics have features in common, differing only in the degree of effects.

DRUGS COMMONLY ABUSED

3. The narcotic receptor cells have been identified in the brain.
   a. Binding to these cells inhibits the transmission of nerve impulses carrying pain signals to the brain from the body.
   b. Evidence also suggests that there may be receptor cells in other areas of the body, which may explain some of the side effects of narcotics.
4. Hormones from the pituitary gland named endorphin and enkephalins, similar in chemical composition to opiates, have been found.
   a. Endorphins and enkephalins seem to be released from the pituitary gland and bind to the narcotics-receptor cells whenever we experience pain or stress.
   b. The amounts of endorphins and enkephalins produced in the body give individuals their differing thresholds to pain, anxiety, and stress.
   c. The use of narcotics and subsequent addiction may be greatly influenced by an individual's level of endorphin, enkephalins, or the amount of pain and stress in his or her environment.
5. Individuals habitually use narcotics to:
   a. Increase their ability to withstand the pain and stress of their environment.
   b. Experience euphoria, relief from fear and apprehension, and feelings of peace and tranquility.
6. Prolonged use of narcotics produces all the classical symptoms of drug addiction.
   a. Abrupt withdrawal causes *narcotic-solvent abstinence syndrome.*
   b. A dependent user is clearly an addict.

C. Effects of abusive dosages
   1. Recognition of narcotics addicts is extremely difficult.
      a. Pupils constrict to a pinpoint condition and do not react to light.
      b. Long-term addicts tend to be pale and emaciated and suffer from severe constipation.
      c. They show little or no interest in sex.
      d. Their appetites are poor.
   2. After the initial "flash" of euphoria, addicts become sleepy, lethargic, semistuporous, and dreamy.
   3. A large percentage of burglary, shoplifting, prostitution, and other petty crime is related to the addict's need for money.
   4. Overdose deaths occur from medullary paralysis by depres-

SUMMARY

sion of the respiratory center in the brain to a degree that the individual stops breathing.
D. Administration of narcotics. Narcotics are administrated either:
   1. Orally
   2. By injection
      a. A subcutaneous or intramuscular injection is called a "skin pop."
      b. Injection into the blood stream through a vein is known as a "mainliner" or "fix."
E. Drugs classified as narcotics
   1. Opium is derived from *Papaver somniferum*. It is seldom used by American addicts who use its derivatives:
      a. Morphine
      b. Heroin
      c. Codeine
   2. Morphine
      a. Morphine is the chief derivative of opium.
      b. When morphine is unlawfully possessed, it usually has been stolen from a physician or pharmacist or obtained by means of forgery on a prescription form stolen from a physician.
   3. Heroin
      a. Heroin is produced from morphine.
      b. Because of its strength, the peddler is able to adulterate it many times before it is sold.
         (1) Milk-sugar, mannite, procaine, quinine and often unsafe chemicals are used to adulterate heroin.
         (2) Street heroin is often 4, 3, or even 1 percent or less in purity.
      c. Usually heroin is injected.
      d. Legally, heroin is considered the most dangerous of all drugs, and its manufacture, sale, and possession is prohibited by federal law.
   4. Codeine
      a. Codeine is milder than other opiates discussed in this section.
      b. Heroin users will sometimes resort to the use of codeine when deprived of their regular supply of heroin.
   5. Synthetic narcotics
      a. These differ from the opiates in that they are made in the laboratory.
      b. Addiction to the synthetics is less likely than addiction to morphine or heroin, but it is possible with all synthetics.

c. Abuse of synthetically produced narcotics is limited mainly to the medical and allied medical professions.
IV. Volatile liquids (easily vaporized chemicals)
   A. Solvents
      1. Inhalation produces a state of intoxication.
      2. Several dangerous solvents are used in the manufacture of airplane glue and cement.
      3. Solvents have a depressant action on the central nervous system that is similar to the action of barbiturates and narcotics.
      4. The abuser experiences effects closely simulating the early stages of alcohol intoxication.
         a. Tolerance develops quickly, and many feel that dependent users are addicted to solvents.
         b. However, upon withdrawal from the solvents, a user experiences few severe withdrawal symptoms.
   B. Amyl nitrite is a volatile liquid sold in ampules covered with a nylon net so that they can be broken and inhaled.
   C. Butyl nitrite is sold as a "room odorizer" and is inhaled.
V. Hypnotic-sedatives
   A. They depress the central nervous system into a condition resembling normal sleep.
      1. All these drugs resemble each other in this action regardless of their chemical differences.
      2. The only difference between a hypnotic action and a sedative action is the degree of depression.
      3. When abused, they greatly reduce anxiety and produce a mild euphoria as a relief for people who tend to live in an uncomfortable, painful, or hyperaroused state.
   B. Barbiturates
      1. They are the most widely used hypnotic-sedatives.
      2. Actions of barbiturates
         a. In low and moderate dosages barbiturates interfere with oxygen consumption and the mechanisms by which energy is derived, stored, and utilized within the cells of the brain.
         b. In higher dosages, barbiturates depress the synapses of neurons bringing information into the reticular formation, reducing wakefulness and alertness.
         c. Some barbiturates work only on the cerebral motor areas, enabling these drugs to selectively prevent epileptic seizures.
         d. At extremely high (toxic, abusive) dosages, barbiturates

SUMMARY

produce a state of hyperactivity and excitement in the reticular activating system before the general suppression of the nervous system sets in.
3. Effects of barbiturates
   a. These pills account for the majority of drug poisoning, most cases of which are considered suicide attempts.
   b. Physicians constantly warn barbiturate-using (sleeping pill) patients not to keep their bottles of tablets near the bed, where they may stretch out a hand to take more pills while in a confused or sleeplike state of mind.
   c. Barbiturates are usually taken orally.
   d. A person under the influence acts like someone who has had enough alcohol to show signs of intoxication, but takes much longer to sober up.
   e. Small abusive doses make an individual feel relaxed, sociable, and good-humored. As dosages are increased, the individual becomes antisocial and quarrelsome and gradually slumps into a deep sleep.
   f. Those who become addicted to barbiturates must have the drugs in order to prevent going into withdrawal (*alcohol-sedative withdrawal syndrome*).
C. Alcohol and alcohol derivatives
   1. The oldest hypnotic-sedatives used in modern medicine were alcohol and its major derivatives.
      a. Alcoholism is such a serious problem that two complete chapters (Chapters 6 and 7) of this book are devoted to it.
      b. *Chloral hydrate* is highly variable in its effects and is extremely dangerous to administer.
      c. *Paraldehyde* is seldom used as an ordinary sleeping drug because of its disagreeable taste and odor.
D. Bromides
   1. They exert a sedative effect on the central nervous system.
   2. They are rarely abused because they do not produce euphoria.
   3. These drugs are highly dangerous because of their extreme toxicity.
E. Other nonbarbiturate hypnotic-sedatives
   1. *Ethchlorvynol* is a mild sedative.
   2. *Ethinamate* is a mild sedative related to tranquilizers.
   3. *Glutethimide* (Doriden or "Ciba") is a hypnotic-sedative that produces effects similar to those of the shorter-acting barbiturates.
   4. *Methyprylon* is similar to the barbiturates.

VI. Antihistamines
   A. Antihistamines act by preventing the actions of histamine.
      1. Histamine stimulates the smooth muscles and is a powerful vasodilator in the capillary beds.
   B. Many antihistamines produce drowsiness that can progress to sedation and produce sleep when used in increasing dosages.

VII. Tranquilizers
   A. The major tranquilizers
      1. Major tranquilizers are used for the management of the severely mentally ill.
      2. These drugs are powerful *antipsychotic agents.*
   B. The minor tranquilizers
      1. Minor tranquilizers are the most widely used group of psychotropic drugs in the United States.
      2. These drugs act directly on the reticular formation and lower areas of the brain.
      3. In the addiction-prone personality, the progressive self-administration of tranquilizers is a very real danger and is considered a form of drug abuse.

VIII. Cannabis (marijuana)
   A. These drugs are produced from the many varieties of *Cannabis sativa* grown throughout the world.
      1. This drug family, more than any other, cannot be accurately explained without specifying dosage levels.
         a. The marijuana used in the United States is probably the weakest preparation of the plant used in the world.
         b. The most potent is *charas,* which is the unadulterated resin obtained from the female plant or its dried flowers.
      2. In large doses, cannabis bears many similarities to the hallucinogenic drugs and to sedatives, alcohol, and narcotics.
   B. Effects of cannabis:
      1. Cannabis intoxication ends in sedation and sleep, whereas wakefulness is characteristic throughout intoxication with hallucinogenic drugs.
      2. Both alcohol and marijuana produce an early excitement and a later sedated phase, but marijuana intoxication is distinguishable by the user from alcoholic intoxication.
      3. The total effects of a marijuana experience last from three to five hours, after which the user feels slightly drowsy or sleepy and hungry.
         a. All behavioral modifications depend upon environment, personal feelings, and the amount of marijuana smoked.

SUMMARY

    b. Cannabis has been shown to stimulate frontal areas of the cortex while depressing parietal areas.
    c. During cannabis intoxication the ability to arouse the cortex is diminished, which suggests further actions within the integrative centers of the hypothalamus, the limbic system, and the reticular activating system.
    d. Thus, it appears that the actions of cannabis in the brain are widespread and related to increasing and decreasing concentrations of cannabis and its metabolic by-products in specific brain areas.

IX. Hallucinogens
  A. A drug that creates vivid distortions in the senses without greatly disturbing the user's consciousness is called a *hallucinogenic* drug.
  B. The actions of hallucinogenic drugs are still rather fragmentary.
    1. Authorities feel that the action is related to serotonin.
    2. One theory is that hallucinogenic drugs either block the production of serotinin, or replace serotinin, flooding the brain stem with these serotoninlike substances.
  C. Phylethylamine drug family (mescaline)
    1. The most significant drug abused in this group is mescaline.
    2. Mescaline is of interest because it is chemically related to the amphetamine compounds and to norepinephrine, which seems to account for its actions on the body.
  D. Indole alkaloids
    1. Probably the most potent indole alkaloids are the ergot alkaloids, which can be changed into LSD.
    2. Following the initial physical effects of the drug, users show a great degree of subjective euphoria and a sense of mental clarity or comprehension.
    3. The temporary modifications within the nervous system caused by LSD may produce any type of behavioral disturbance.
    4. Because of the large number of extremely disturbed reactions that are occurring with the illegal use of LSD, it is possible that much of the illegal supplies of LSD are either contaminated or mixed with other substances.

X. Cocaine
  A. Cocaine is the principal active ingredient of the South American coca plant.
  B. Processed cocaine is an odorless, white, fluffy, fine crystalline powder, similar to snow in appearance.

1. Stimulating effects on the cerebral cortex result in euphoria, laughter, restlessness, and excitement.
   C. There are two distinct types of cocaine poisoning:
      1. In the first, death is due to circulatory collapse.
         a. This is caused by a drug idiosyncrasy.
         b. There is no way to tell when or where this will take place.
      2. The second type is due to respiratory failure from an overdose of the drug.
XI. Antidepressants
   A. The actions of these drugs on the central nervous system are based on what is known as the *amine theory of mood*.
      1. These substances are known chemically as monoamines because they contain a single amino group in their chemical structure.
      2. The actions of many of the previously discussed drug groups seem to fit this theory.
   B. Psychic energizers—These are mild cerebral stimulants that produce an elevation of mood, increased activity, heightened confidence, and increased ability to concentrate.
   C. Antidepressants—These drugs make a person more cheerful, active, and interested and also relieve hopelessness, despair, fatigue, and guilt to a greater degree than do psychic energizers.
   D. Anticholinergics—These drugs make a person restless, wakeful, and talkative.
XII. Amphetamines
   A. These comprise a large group of synthetic drugs that stimulate the central nervous system, particularly the reticular formation and the cerebral cortex.
   B. The primary action of amphetamines is increasing the release and blocking the deactivation of catecholamines.
   C. In 1970 the legal use of amphetamines was restricted to three types of conditions:
      1. Short-term weight-reducing programs
      2. Narcolepsy
      3. Hyperkinetic or hyperactive children
   D. Abuse of amphetamines throws an individual into the *Speed Cycle*.
XIII. Poly drug use—Instead of a single drug being used as a recreational drug, it is common to find that many substances are being taken simultaneously or in sequence.

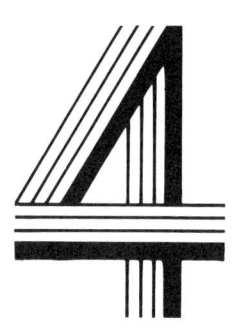

PSYCHOLOGICAL AND SOCIOLOGICAL FACTORS IN DRUG ABUSE

Since the use of drugs in medical practice is accepted as legitimate, the phrases "drug use," "drug user," and "recreational user," usually refer to the abuse of some substance, that is, the use of drugs for purposes other than medicinal. Such abuse is usually associated with dosages (abusive dosages) many times in excess of those prescribed in legitimate medical practice. When a drug (or substance) is self-administered in toxic dosages and damages an individual, a society, or both, the drug (or substance) is being abused.

Theoretically, any drug can be abused. Despite the usefulness of the foregoing definition, drug abuse is still a difficult concept to define completely and adequately. The continuum of responsibility from individual to society frequently clouds important scientific and medical debates about drugs. Attitudes toward drug abuse have taken on political significance, again adding to the scientific confusion.

PATTERNS OF DRUG ABUSE

In most cases of repeated abuse of a mood-modifying drug, an individual develops a dependency upon either the psychological effects of the substance or the social patterns associated with its abuse. Some substances, such as nicotine (in tobacco) and caffeine (in coffee and tea), cause some stimulation and a degree of dependence in the user but are socially acceptable because they do not produce effects such as euphoria and other extreme mood modifications. Alcohol may produce such effects to some degree, but in moderate amounts it is still socially acceptable.

Drugs that are considered dangerous and unacceptable to society, when used for other than legitimate medical purposes, are those substances which create problems within society by producing personality changes, euphoria, or abnormal social behavior in users. Drugs so classified include narcotics, solvents, hypnotics, sedatives, tranquilizers, *Cannabis* derivatives, hallucinogens, cocaine, and amphetamines. They may be called mood modifiers, psychotropic drugs, or psychotoxic drugs. These terms describe the effects of a drug and are not scientific classifications.

Effects are not produced until drugs interact biologically, psychologically, and socially with the individual. Changes in moods, perceptions, feelings, and behavior are interactions involving the drug being abused, the way it enters the body (inhalation, oral ingestion, or injection), the amount present (minimal to lethal dosage), the speed of reaction (position of the drug on the continuum), and the social setting in which the drug is taken.

Drug Behavior

Not everyone who explores the effects of mood-modifying drugs will follow the same predictable pattern of behavior. Some individuals are introduced to drugs through social influences; many, through curiosity about the effects; others, because of emotional problems.

Experimenters. Well over half of all individuals who are reported to be drug users turn out to have used drugs only a few times and have no intentions of using illicit mood-modifying drugs again. Such people, in reality, are drug experimenters or tasters. They have little place in a discussion of drug abuse because they have already made their decision regarding the abuse of drugs. But too often, they have been arrested during their experimentation and may carry a criminal record of their experiment for the rest of their lives.

Occasional users. Many experimenters find they need the social or personal gratification that mood-modifying drugs can give them. They continue their experimentation with drugs and may be referred to as occasional users. These individuals are often socially conscious and use the current "in" drug or drugs. When this behavior pattern occurs with the use of alcohol, these individuals are called "social drinkers" or "recreational users." Only a fine line divides this occasional user from the regular user.

Regular users. Some people abuse drugs consistently. These individuals are regular users. Such individuals use drugs regularly (one or more times per week), and they often experiment with a variety of different drugs and still use the "in" drugs. They function within society, but drug abuse is the central focus of their lives. It is not just an intermittent assist in the pursuit of life but part of a drug-oriented ideology and a membership card in a life-style or subculture.

These individuals do not seek help from clinics or consulting rooms, and the physician may never see them. They usually recover from their drug experiences, none the worse for wear, often proclaiming loudly that they have gained a profound and valuable insight into themselves, nature, religion, and so forth. Such individuals are often the greatest defenders of the personal right to use drugs. They have very closed minds concerning drugs because of street experimentation: "I know because I have been there." Judged by common criteria of mental health—the ability to work, to love, and to function within a society—they seem neither less nor more effective after their drug experiences. The most dangerous aspect of regular drug use is the danger of lapsing into drug dependency.

Drug Dependency

Dependency on drugs is often associated with an inadequate personality. Whether this personality disorder is the cause or the result of the drug abuse is not always apparent. The importance of predisposing hereditary physiological factors has also not been determined. But the abnormal social behavior, the modification of moods, and the personality problems of such individuals are more of a problem to society than is the drug abuse itself. Such actions label this individual as being emotionally disturbed.

Researchers in the field of drugs have traditionally attempted to distinguish forms of drug dependence according to the nature of the drug used. The drugs have been classified as addicting, habituating, or habit forming. Although these terms may be useful in describing the effects of certain drugs, when considering the individual, it is the drug dependence itself, whatever form it takes, that is significant. However, the terms "habituation" and "addiction" have been redefined and enlarged in the hope that they can still serve the needs not only of science and medicine but also of law and society.

The definitions of "habituation" and "addiction" most often quoted today are those of the Expert Committee on Addiction-Producing Drugs of the World Health Organization. Table 4.1 compares the differences and similarities between the two terms.

Drug habituation.   A person said to have a "drug habit" seeks out the drug and has a desire, although not an overriding compulsion, to con-

TABLE 4.1
Definitions of Addiction and Habituation

Drug Addiction	Drug Habituation
Drug addiction is a state of periodic or chronic intoxication produced by the repeated consumption of a drug (natural or synthetic). Characteristics:	Drug habituation is a condition resulting from the repeated consumption of a drug. Characteristics:
1. an overpowering desire or need (compulsion) to continue taking the drug and to obtain it by any means	1. a desire (but not a compulsion) to continue taking the drug for the sense of improved well-being or effect it produces
2. a tendency to increase the dose	2. little or no tendency to increase the dose
3. a psychic (psychological) and generally a physical dependence on the effects of the drug	3. some degree of psychic dependence on the effect of the drug, but absence of physical dependence and hence of abstinence syndrome
4. a detrimental effect on the individual and on society	4. detrimental effects, if any, primarily on the individual

Source: Modified from Maurice H. Seevers, "Medical Perspectives on Habituation and Addiction," *Journal of the American Medical Association,* 181, No. 2 (July 14, 1962), 92–98, Table 93.

tinue taking the drug. Consequently, there is some degree of psychic dependence upon the use of the drug. This can produce both physical and psychological effects that are detrimental to the individual.

The physical properties of a habituating drug must be considered. Habituating drugs do not produce physical dependence or withdrawal symptoms (abstinence syndrome) when a person suddenly stops using them. Habituating drugs may or may not produce a tendency to increase the dosage (tolerance) for the desired euphoric, mood-modifying, or behavior-changing effects.

Tolerance is a condition shared by some habituating and all addicting drugs. Tolerance is a fundamental survival mechanism within the cells of the body that permits them to be exposed continuously to poisonous substances in toxic dosages without evoking dangerous (or deadly) responses. It is the cumulative resistance to the effects of some drugs. Tolerance has begun to develop when repeated administration of a given dosage of a drug produces a decreasing effect and increasingly higher dosages must be taken in order to obtain the original effects.

Medically, individuals abusing habituating drugs are called "drug users" or "users."

Drug addiction. In addition to causing psychological dependence, some drugs evoke biochemical and physiological adaptions in the user known as tolerance and physical dependence.

As shown in Table 4.1, the World Health Organization has defined addiction as:

> a state of periodic or chronic intoxication produced by the repeated consumption of a drug (natural or synthetic), which produces the following characteristics: (1) an overpowering desire or compulsion to continue taking the drug and to obtain it by any means: (2) a tendency to increase the dosage, showing body tolerance; (3) a psychic and generally a physical dependence on the effects of the drug; and (4) the creation of an individual and social problem.

With the chronic, repeated abuse of addicting drugs, two mutually dependent conditions, physical dependence and withdrawal illness, develop.

Physical dependence allows the drug-tolerant cells to continue functioning in a way that is similar to the way they functioned before the drug was first introduced into the body. As physical dependence develops, the mood-modifying responses a drug abuser is seeking decrease in intensity. This either causes the individual to continue to increase dosages until he or she reaches a lethal dosage, "ODs," and dies or is forced to go through withdrawal, allowing the cells to return to a normal, pretolerant state. The existence of physical dependence is shown as a clearly defined syndrome of increased nervous activity evident to an individual only when administration of the drug is stopped.

Withdrawal illness, or abstinence syndrome, is clearly defined and exists in two distinct types. *Narcotic-solvent* abstinence syndrome is produced by the sudden withdrawal of narcotics or solvents. The symptoms include irritability, emotional depression, extreme nervousness, pain in the abdomen, and nausea, all of which may range from mild to severe. Sudden withdrawal from minor tranquilizers, alcohol, or barbiturates produces *alcohol-barbiturate* abstinence syndrome. Alcohol withdrawal evokes a mild-to-severe form known as *delirium tremens.* The sudden withdrawal of barbiturates from a person with physical dependence may cause convulsions, along with the delirium tremens, frequently resulting in death. No clear physical dependence occurs with any of the stimulant drugs; consequently, upon discontinuing their use, there may be extreme fatigue and depression, but no true abstinence syndrome is produced.

Addict or alcoholic. The word "addict" is applied to someone chronically abusing physically dependent, or addicting, drugs. These are the depressant drugs such as opium and its derivatives, synthetic narcotics, barbiturates, and solvents. The word "alcoholic" is applied to someone who is abusing alcoholic beverages.

Development of Drug Dependency

In all cases of repeated drug abuse, the person who misuses such drugs voluntarily chooses to do so. Once started, such abuse of any mood-modifying drug will usually lead to a drug dependence. The effect of the drug dependence upon the individual is related to the nature of the mood-modifying drug, the quantity used, and the emotional, social, and environmental characteristics of the individual.

During the early stages of drug abuse, the major force inducing drug-seeking behavior, producing psychological dependence, is the reward of a pleasurable experience. Also, many sociological conditioning factors in the drug abuser's environment are involved in drug dependence.

The person who does not find a pleasurable experience in drug use or who is subjected to strong negative social pressures when first experimenting with a drug does not usually continue to abuse drugs. If his or her personality does not lack or seek what drugs have to offer, or if drug use is not socially reinforced, that person is not likely to continue drug use.

If the first drug trial is a rewarding experience emotionally or socially, a few more rewarding experiments may follow, until drug-taking becomes a conditioned pattern of behavior. Continued positive reinforcement (rewarding experiences) the use of drugs leads in time to a psychological dependence. This drug dependence is all that is needed to lead susceptible individuals to dependent drug abuse.

Thrill-seeking and the pursuit of new or novel experiences seem to be factors that apply to all drugs, stimulants, and depressants alike, and one basis of psychological dependence resulting from the positive reinforcement of reward.

Upon attempting to discontinue use of one of the addicting drugs, the drug-dependent individual will experience withdrawal. Having once suffered these adverse symptoms, the addict will also be negatively reinforced to seek addicting drugs in order to avoid abstinence syndrome. The conditioning pattern of positive reward and negative withdrawal establishes the strong psychological dependence of addicting drugs. Often, avoiding the negative properties becomes more important than the reward. The individual becomes very antisocial, and his or her life is little more than a prolonged criminal experience of petty thievery or prostitution to obtain money for the drug being abused in order to avoid the adverse properties of withdrawal.

The reward of depressant drugs is their ability to reduce mental and physical functioning, enabling the individual to escape from pain, anxiety, depression, fatigue, aggressive feelings, or even excessive passivity or boredom.

With the stimulant-dependent person, the reward is associated with the direct effects of the drug. The predominant reward actions of the stimulant drugs are their abilities to enhance the psychological functions and increase motor activity (increased capacity to perform physically). Commonly, this broad group of drugs (Figure 2.9) induces distortions of perception, hallucinations, illusions, and drastic disorders of behavior often similar to states of mental illness. The dependence upon these drugs and the tendency to increase dosages is not as much related to physical tolerance as it is to a desire for increased rewards.

During prolonged drug dependence, tolerance and increased reward-seeking diminish the intensity of drug reward, causing the individual to increase the amount of drug being taken, to mix different drugs, or to progress up or down the drug continuum (Figure 2.9) in search of stronger and more intense drugs.

New Look at Drug Dependency

Much research has dealt with the harmful effects—physical, psychological, and social—of substance abuse. And certainly many damaging results are known. But too little attention has been given to the *functions of drug dependency*. For there are such functions, and they are important in that they provide the motivation for continued drug abuse. In concentrating our attention on the negative aspects of drug dependency, we have often failed to notice the functions or roles that the abused substance or even the dependency itself plays in the life of the dependent individual. We must understand not only what the substance does *to* the user, but also what it does *for* him or her. Only when dependency is viewed in terms of its function to the individual can efforts toward eliminating the dependency succeed. Seldom if ever can the problem be solved by merely trying to keep a person from abusing a drug if nothing is provided to more effectively fulfill the functions that the drug has served.

Let's start with an overview of some of the functions of dependency. (Most of these ideas will be expanded later in this chapter.) In the first place, substance dependency is a way of life. It is a way of dealing with difficulty; of soothing life's pains. It can provide escape from tension, fear, anger, feelings of inadequacy, or from the daily insults of poverty and oppression. Dependency is a means of achieving in fantasy what cannot be attained in reality. It is a means of being (in fantasy) whatever you want to be: braver, smarter, better looking, more likeable, more competent. It is a coping mechanism in a world that is too difficult or too alien. It expresses anger and represses anger. One can live with rage for a long time by using enough of some substance. Dependency rescues people from the depths of loneliness and isolation.

Socially, dependency can provide individual and group identity. A drug user automatically joins a worldwide club. There are many others like him or her, and they have a great deal to share within a whole subculture. For many people, this is the first experience of a sense of belonging.

Although we tend to see substance abuse as negative and destructive we must not forget the meanings, values, and many functions of the substance and the dependency in the life of the dependent person. Substituting other more sustaining and fulfilling values and meanings to life is essential if people are going to be able to let go of substance dependencies and create lives that are both possible and satisfying without drugs.

PSYCHOLOGICAL FACTORS IN DRUG ABUSE

There is little argument that drug use occupies a prominent position in the total spectrum of emotional health. It can be seen as causing emotional health problems, as being symptomatic of emotional disturbances, and, under medical direction, as playing a major role in the treatment of such problems. Psychoactive drugs are what most people would call a "mixed blessing."

The concept of emotional health may be approached from many directions. A "model" of emotional health that may contribute to an understanding of drug abuse and dependency is as follows:

1. Every person has certain basic needs.
2. One element of emotional health is the development of effective, appropriate, and socially acceptable methods of fulfilling these needs.
3. However, even having accomplished this, there is always an unavoidable amount of frustration of a person's needs. No one can have all of his or her needs fulfilled all of the time.
4. Thus, another element of emotional health is the development of effective, appropriate, and socially acceptable methods of dealing with the frustration of your needs. These may be called "coping devices" or, more traditionally, "ego-defense mechanisms."
5. The excessive use of alcohol or other drugs by many people is viewed as an inappropriate method of attempting to fulfill needs or of dealing with the feelings of frustration resulting from unfulfilled needs.
6. The rehabilitation of many drug-dependent persons can only come about through their development of more adequate need-fulfillment techniques and/or improved methods of coping with need-frustration.

Personality

Personality is the total reaction of a person to his or her environment. It includes how an individual reacts to any particular situation he or she encounters. It incorporates traits, behavior patterns, and attitudes established over the years of one's life. Everything that has ever happened to you has produced some effect upon your personality. This includes how you look at your surroundings and how you learn to deal with your basic emotional needs and the obstacles presented by the world around you. Also, your personality is not static but dynamic, constantly changing as you go through life.

At least four forces influence personality: genetic makeup, physical and social environment, interpretation of the environment, and the learned ways of coping (coping behavior). The first two are largely beyond individual control. You have no control over your hereditary traits, and personality traits are to some extent hereditary. You have a limited ability to influence environment (where you live). But your interpretation of environment and methods of coping are under your direct control and are the basis of your ability to destroy or to improve your personality and behavior.

Personality Structure

Many attempts have been made to describe the factors controlling personality. For the purposes of this discussion, the concepts of pioneer psychoanalyst Sigmund Freud serve to describe the control of personality. Freud conceived personality as being divided into three "control processes": the id, the ego, and the superego. Freud felt that the mind functioned at two levels: the conscious mind, which governs the observable actions of the individual, and the unconscious mind, which governs all the unrealized reasons, instincts, impressions, and emotional mechanisms of control within an individual. He envisioned the personality and its control as these processes working within the two levels of consciousness.

The id processes. At the time of birth each of us carries a bundle of inborn tendencies necessary for survival. These inborn tendencies represent the unconscious strivings of the human to live (survive) and *enjoy* life. We quickly learn what brings us pleasure. Since the id processes work for the individual's biological survival and pleasure, they develop early and are very aggressive and selfish. They operate for our pleasure and gratification and do not distinguish between good and evil or between what is realistic and what is not. These processes operate under the pleasure principle ("whatever brings pleasure is good"), and if someone is going to control these processes within the framework of society, he or she must learn how to tame the id.

The ego processes. The ego represents the conscious mind. It acts as a moderator between the id and the outer world. It is in contact with the environment (reality) and with the id and the superego (sense of right and wrong). The ego must integrate all factors and determine appropriate behavior for each situation. The ego must decide which urges can be allowed satisfaction and which urges must be suppressed. Since the individual at birth lacks training with respect to such responses, the processes of the ego develop gradually and gain strength as the individual grows older and responds to the pressures of society.

The superego processes. The superego represents the judgment mechanisms regarding right and wrong, good and evil. It advises and threatens the ego. As very young children we have little concept of what is right and wrong, but as the codes and values of society are impressed upon us, we gradually develop our superegos. The main basis of the superegos of children is their concept of their parents. Children use their parents as a guide for their own behavior. Other adults, peers, and the values and laws of their society also contribute to the superego. The superego is thought to lie in both the conscious and unconscious portions of the mind. The conscious portion of the superego is the "conscience." Behavior that violates the judgment of the superego produces guilt. Guilt is an extremely uncomfortable feeling. The need to reduce this feeling is one of the factors motivating people to abuse the "escape" drugs such as alcohol, opiates, and barbiturate depressants.

Basic Human Needs

In order to strengthen your personality, you must know yourself. Why do you do the things you do? Why do you feel the way you do? The answers to these questions will come largely from your understanding of your basic needs, how these needs are fulfilled, and how you react to the frustration of these needs. The basic human needs follow a sequence from the most basic need to the higher needs.[1] The higher needs develop only when the basic needs are fulfilled:

1. Physiological needs
2. Safety and security needs
3. Need for love and belonging
4. Esteem needs
5. Need for self-actualization
6. Need to know and understand
7. Aesthetic needs

Among these, the following seem to be the ones more closely related to patterns of drug abuse.

---

[1] From A. H. Maslow, *Motivation and Personality,* 2nd ed. (New York: Harper & Row, 1970).

Physiological needs. The most basic human needs are the physical drives such as hunger, thirst, sleep, and sexual aggressiveness. When these are unsatisfied, all other needs are pushed into the background. For example, to an extremely hungry person, utopia is merely a place where there is plenty of food. "If I were guaranteed food (or heroin or a replacement such as methadone) for the rest of my life, I would never want anything more." But this is not so; when physiological needs are satisfied, other, higher needs immediately emerge and begin to dominate the person. As long as a need is satisfied (the rationale for methadone use), that need has little effect on behavior. Behavior is governed more by unsatisfied needs, which, beyond these physiological needs, are all emotional.

Security and safety needs. Once physiological needs are fairly well satisfied, the security needs emerge. Security is produced by stability, consistency, and knowledge of the limits of permissiveness. These should be established by parents through a consistent framework of rules of conduct; but if they are not established for an individual, that person must produce his or her own in order to mature. Often, in the case of drug abuse, once a person is able to formulate his or her own code of conduct, the drugs become less important and the person's security is established.

Need for love and belonging. Belonging is a very important factor in the first experimentations with drugs. Young people feel a strong need for friends, companionship, acceptance by a group. If these are tied to drug abuse and if immediate social satisfaction is derived from the group, other possible benefits (security, esteem, and self-fulfillment) will cause the individual to become further involved in a drug subculture.

But rejection by a group may also cause an individual to withdraw with drugs because of fear of further rejection. These feelings often lead an individual to depressant drugs because of the physical effects (calming, euphoria, and feelings of security) while under their influence.

Opportunities for group involvement without drugs help an individual to keep from becoming involved with drugs. Belonging to a group also helps individuals reduce their involvement with drugs. Such belonging is the basis of therapeutic communities such as Synanon, Daytop Village, Odyssey House, and Gateway House.

Esteem needs. Everyone needs self-esteem—a feeling of personal worth or value. The daily lives of a great many of us, perhaps most of us, are greatly influenced by our efforts to fulfill this need. Self-esteem is one of the hardest needs to satisfy in an alcoholic or other drug abuser. A "vicious circle" may be set up in which lack of self-esteem motivates drug dependency, while the drug dependency causes a further loss of self-esteem. Until something intervenes to break this cycle of hopelessness, drug abuse is likely to continue. For many people, the best way to start building self-esteem is to find some activity in which success is attainable. This could be in a job, a hobby, a sport, or in service to others. Activities are especially helpful when they combine a sense of personal achievement with acceptability to other people.

Need for self-actualization. The need for self-actualization is merely the need to do what you are capable of doing—to make full use of your abilities in order to feel fulfilled as a person. Very few people, according to A. H. Maslow, are fully self-actualized. Several forces may act to block such fulfillment. Many of us, for example, are still trying to fulfill needs which arise earlier in the sequence of needs, such as the need for a feeling of self-esteem. Others of us may not be aware of our full potential for achievement. Still others find that forces beyond their control, such as the social structure, prejudices, or economic conditions, act to prevent them from making full use of their abilities. Many people feel a great sense of frustration in being blocked from doing what they know they are capable of doing. This feeling may be so overwhelming that it motivates a drug- or alcohol-induced retreat from reality.

Need to know and understand. This need, when fulfilled, satisfies curiosity. Curiosity is a natural rather than a learned characteristic. A young child is naturally curious, and in the early grades a good educator will try, very carefully, not to destroy this valuable trait. The first experimentation in drugs is often an expression of curiosity, too often, without prior knowledge of the consequences of drug use. This experimentation can produce the thrills curious young people are looking for in an extremely structured society. Drug education can be instrumental in reducing drug experimentation.

Emotional Dysorganization

The ego tries to level off all emotional disturbances as best it can. At times, however, each of us experiences frustration of our efforts to fulfill our basic needs. Aggressive impulses of the id, over which the ego has maintained control, are aroused and threaten that control. From the other side, the superego's ideal may be impossible to attain during these periods, so the ego, which tries to regulate the self with regard to the possible, gives way. As this continues, over a period of time, the ability of the ego processes to maintain normal control diminishes, producing what is commonly called emotional, or mental, illness. (The two terms are synonymous.)

The late Dr. Karl Menninger, co-founder of the Menninger clinic, a psychiatric hospital in Topeka, Kansas, regards emotional illness as a

PSYCHOLOGICAL FACTORS IN DRUG ABUSE

continuum that ranges from emotional maturity or relative normality to emotional collapse or death. He coined the word "dysorganization" to describe the increasingly painful or disturbing experiences the ego goes through in trying to maintain successful control over the id processes. The prefix *dys-* means "difficult"; thus, this word refers to the disturbance and pain the person experiences in trying to maintain an organized emotional balance. (Do not mistake this word for "disorganization," which means the state of being disarranged.) In dysorganization, emotional organization is made difficult; it is not necessarily lost or destroyed.

According to Dr. Menninger there are five levels of emotional dysorganization, ranging from mild to severe (Figure 4.1). These are

NORMAL EMOTIONAL ORGANIZATION

Diminishing Ego Control

LEVEL 1
Strong tensions, anxieties, nervousness

LEVEL 2
Phobias, personality deformities, addiction, alcoholism, obsessions, neuroses

LEVEL 3
Open aggressions, homicides, attacks, assaults, social offenses

LEVEL 4
Classical severe mental illnesses, paranoias, depressions, delirium, melancholia, psychoses, insanities

Increasing Severity of Dysorganization

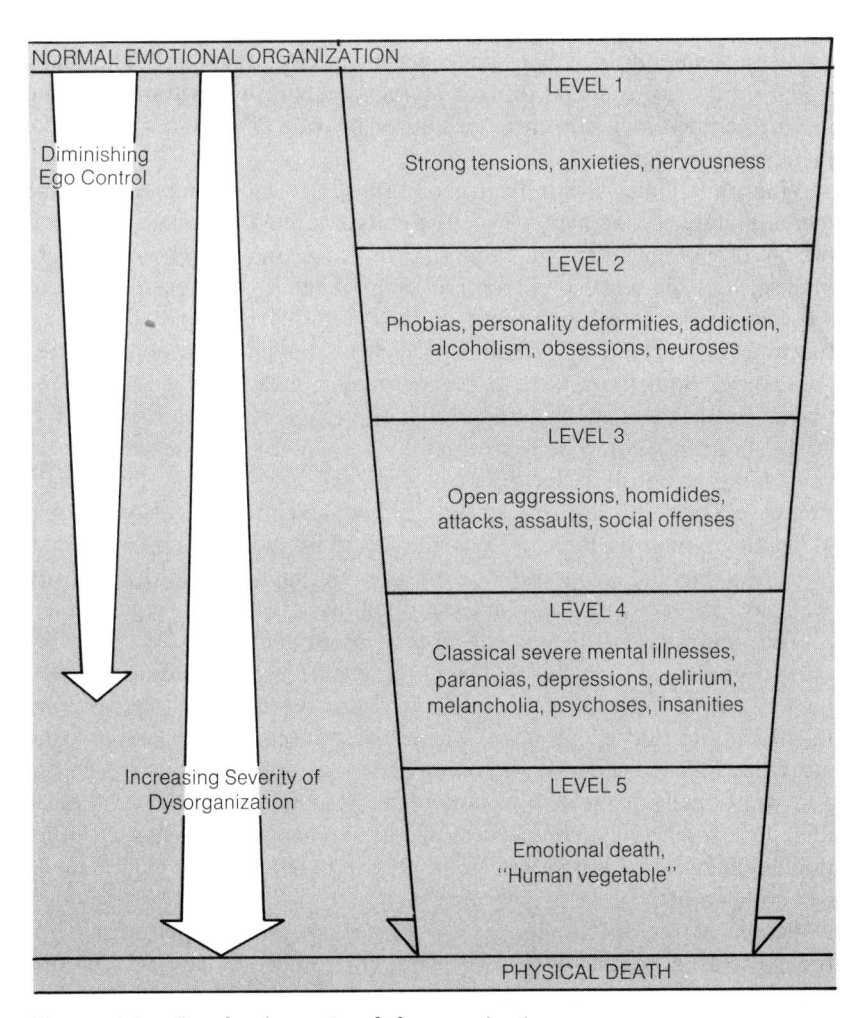

LEVEL 5
Emotional death, "Human vegetable"

PHYSICAL DEATH

Figure 4.1. Levels of emotional dysorganization.

points on a continuum, and many individuals fall between described levels. Starting with a relatively normal adjustment and ego control, Dr. Menninger describes increasing degrees of emotional dysorganization that represent degrees of mental illness. As the pressure on the ego becomes greater, the individual's whole self becomes more and more threatened. At the extreme, the person may even die because the whole system, both emotional and physical, may collapse under the unbearable strain of dysorganization.

The person entering the continuum of emotional dysorganization experiences somewhat more than the normal amount of tension and shows it through nervousness. At the *first level of dysorganization,* the ego is aware of an increase in aggressive impulses (fear, anger, frustration) coming from the id. Although the person may not recognize these feelings as threats to the ego, there is an awareness of using more than an average amount of willpower to master internal reactions. The occasional drug user or social drinker starts to establish a pattern of using drugs to either block out these feelings or provide experiences that make them seem insignificant.

When not under the influence of drugs, the ego processes become increasingly alert, or hyperalert, to events around the person in anticipation of possible dangers. Sounds are exaggerated; lights are more intense; and all sensory perception seems keener. The person unconsciously refuses to relax at night during sleep. There may be increased touchiness, tearfulness, irritability, nervous laughter, moodiness, or depression. Sometimes such a person may appear overactive by displaying restlessness: walking the floor, biting fingernails, or driving aimlessly about in a car. The person may worry or think obsessively about something, anxiously reflecting on it time and again, or may daydream excessively and thus impair realistic thinking and effective action. People at this level may be hyperzealous in identifying with a "cause," which often becomes the promotion of drug use. By this time the drug abuser may have moved from being an occasional user to being a regular user.

The symptoms of the *second level of dysorganization* are somewhat more pronounced than those of the first level. Since, according to Menninger, the majority of alcoholic and other drug-dependent people are functioning at the second level, we will describe this level in somewhat more detail. The second-level coping processes are definitely unpleasant and emotionally painful. The individual is unhappy and feels a sense of failure, uselessness, or depression. Other people are seen as either indifferent or as definitely antagonistic, even though they may actually be trying to help.

People at the second level may abuse drugs either periodically or regularly as a means of coping with environmental stresses. During periods between drug use, they may function normally within society. But when stresses are placed upon them, they turn to drugs. The personality changes that occur while under the influence of drugs may be very dramatic.

The compulsive or dependent drug abuser, addict, and alcoholic make drug abuse a permanent part of their personalities. Such people have a personality deformity often referred to as the "addictive personality." In view of the millions of Americans who are addicted to alcohol and narcotics and who chronically abuse other drugs, this category represents the largest single psychiatric affliction in the United States. In spite of the fact that addiction is painful and can disable a person, there is a common misunderstanding among people in general about this kind of illness or disease. It is too often considered simply a vice that the victim has chosen voluntarily. In reality it often represents a chemically induced escape from reality. Thus, it represents an inappropriate method of coping with life—an emotional illness.

The *third level* is characterized by the expression of open and direct aggression toward people and things. The id escapes the control of the ego, and all or nearly all the restricting control has been lost.

At the *fourth level* there is expression of the classical mental disorders. Control of the id has been lost; thought processes are badly disrupted; there is no contact with reality; and the interpretation of the outside world is distorted. Behavior may be bizarre, inappropriate, exaggerated, and unpredictable. The numerous psychotic illnesses are symptoms of fourth-level behavior.

Habitual drug abusers periodically dip into levels three and four, for example, the alcoholic person who beats or abuses members of his or her family or the drunk who can always find a fight. The desperate acts and crimes committed to obtain money for drugs or the extreme depression after prolonged use of stimulant drugs and the psychotic states produced by the hallucinogenic drugs are third- and fourth-level behavior.

The *fifth level* represents the total collapse of a person's coping system. It is a frequent outcome of years of dependency on alcohol or other drugs. The will to live is gone. All that is left is a self-destructive determination to end life or to settle for minimal existence. Drug dependency often represents a person's efforts to escape from reality. The permanent escape of suicide is often appealing to such a person. Also, while actually under the influence of alcohol or other drugs, a person's ability to resist the suicide impulse is weakened.

SOCIOLOGICAL FACTORS IN DRUG ABUSE

The interactions between individual drug abuse and society are complex and, frequently, controversial. On the one hand, drug abuse places a heavy burden on society in terms of its many human and monetary costs; while on the other hand, much drug abuse is motivated by such diverse social factors as oppressive living conditions, lack of economic opportunity, and social acceptance or approval of certain forms of drug use. Either society or the individual may be viewed as the victim (or the victimized) depending on one's point of view and the circumstances surrounding a particular case.

In today's society drugs cut across all socioeconomic strata. While different drugs find favor with different subcultures, no segment of society is immune to drug abuse and dependency. And, of course, alcohol abuse pervades every element of society. For persons of limited coping abilities, any life-style presents stresses or situations that can motivate the abuse of alcohol or other drugs. It's not easy to be poor. It's not easy to be rich. It's not easy to go through life as a "nobody." It's not easy to be a celebrity. The social situations of some people are undeniably more difficult or demanding than those of other people. Yet in all of these diverse conditions, many people cope very successfully without excessive drug use.

Drug use itself is causing changes in the social structure. This is shown by the increasing number of individuals who try mood-modifying drugs at some time, use them occasionally, use drugs regularly, or defend the personal right of those who use drugs to do so.

Because of the lack of understanding of the causes of drug abuse, there are no proven generalizations that can be applied to the entire drug-abusing population. One theory holds that the persistent craving for drugs such as narcotics is the result of metabolic or other physiological changes caused by repeated drug use. Another belief is that drugs are abused by people with poorly developed coping abilities as a means of dealing with their psychological difficulties, by the economically deprived to escape the reality of their limited opportunities, by those who, for other reasons, by repeated experimentation have conditioned themselves to respond to stressful stimuli by additional drug abuse. At the other extreme are those holding to the theory that drugs are abused purely for hedonistic purposes by delinquent, criminal individuals who are completely unconcerned with what society expects and are too immature to worry about possible long-term consequences.

Some element of each of these theories can be identified among the extremely wide variety of drug-abuse cases in the United States. There is no doubt that some patterns of drug abuse are specific to certain social, cultural, and economic segments of the population, for example, the proportion of heroin addicts in urban areas, the majority of marijuana usage among the 20- to 30-year-olds, and the excessive use of barbiturates and amphetamines in the suburbs. These patterns are focusing attention on such areas and prompting researchers to ask very specific cause-and-effect questions. Yet, regardless of these patterns, all drug abuse ultimately depends upon the individual.

People seek comfort with a wide variety of chemicals. Some are commonly called drugs; some are not. Some are illegal; some are legal. Regardless of labels or of legal status, all are drugs, and all act according to the same principles. In order to understand drug use, it is necessary to make distinctions among drugs and among patterns of use.

Drug abuse depends upon the individual's reasons for taking a drug and the changes he or she expects to experience. Social factors not only influence a person's reactions to a specific drug but also are important in determining use or nonuse. How someone preceives his or her own pattern of drug use, how it contributes to or interferes with important goals, friendships, and social contacts often determines the substance used (legal or illegal), the pattern and frequency of use, and the outcome of drug use.

Drug abuse is not a problem that can be solved by its victims alone. It is a social problem that concerns everyone. The abuse of drugs among adolescents and young adults is nothing new. People were reported to be addicted to narcotics as early as 1900, and an increase in the number was noted immediately after World Wars I and II. Addiction of young people to narcotics gradually decreased, reaching a low in 1948. Since then, addiction and the abuse of other drugs has continued to rise.

The assistance of relatives and friends of people experimenting with drugs can be invaluable in helping them escape drug dependence. Like the parents of delinquents, the parents of a young drug user sometimes pay little attention to their youngster's friends habits and the places he or she frequents. Although it is often difficult for parents to recognize drug abuse in their children, there are certain clues that may indicate the development of drug dependency. These are not presented here with the intent of fostering an attitude of distrust or suspicion between parents and children. Parents of children who are slipping into drug dependency can deal with this problem much more effectively if their approach is that of helping their children rather than one of overreaction such as panic, rejection, or extreme punishment. Parents, in their reaction to their children's drug usage, can either open new lines of communication with their children, or they can create barriers that totally eliminate any chance of helping them. The following clues are presented in the form of questions that parents may ask themselves:

1. *Has your child become secretive about friends and activities?* Does he or she seem to be involved in some kind of private business? In making appointments? In meeting people? Does he or she suddenly disappear at times? Are unlikely excuses given for these actions?

2. *Are negative attitudes developing?* Does he or she make fun of parents, teachers, police officers, or "straight" young people? Does he or she make fun of the high standards of people who used to be friends?

3. *Is he or she slipping in school?* Is attendance poor? Are grades getting lower? Has he or she dropped the athletics that they formerly

enjoyed? Has he or she given up clubs and social life? Has interest in the opposite sex lessened?

4. *Has he or she begun to have drowsy spells during the day?* Have there been times when it was difficult to awaken him or her? Has he or she seemed "out of it" or knocked out rather than sleepy?

5. *Are there periods of undue excitement?* Is he or she sometimes walking on air? Does he or she seem drunk? Are there any unusual smells?

6. *Is there an excessive need of money?* Is he or she always hunting jobs, losing jobs, or changing jobs? Is there talk of leaving school to make more money? Are items disappearing from the home? Is money missing from your purse or wallet? Is there little or nothing to show for the money he or she has had?

7. *Are his or her arms kept covered?* Are there attempts to hide needle marks? Are there sores on the skin? Is there the constipation that some drugs produce?

8. *Have you found a "stash"?* A collection of pills or powders? Rolling papers? Other drug-use paraphernalia?

These danger signals are for the recognition of drug abuse, but whether or not an individual chooses to try and then to continue to use drugs depends upon his or her personality, the availability of drugs, and the attitudes either for or against drugs prevailing in his or her immediate and overall social group. Before a person ever comes into contact with drugs, he or she often has an attitude either for or against their use. This attitude, in general, corresponds to that of the person's neighborhood, friends and social group, or the community in which he or she lives. In areas where the use of drugs carries less social approval, both the availability of drugs and the incidence of drug abuse are lower than in other areas.

Equally crucial to the person's continued use of drugs is the reaction of his or her social group to an initial drug experience: approval or disapproval, reward or punishment, praise or ridicule. If a person decides upon the continued use of drugs, two consequences are likely: (1) The drug user will become progressively disillusioned with values and standards of nondrug users and will begin to feel abused and misunderstood by their standards. (2) His or her identification and ties with the standards of the drug world will increase.

Advance knowledge of both the properties of drugs and the social reactions to their abuse can influence the course of an individual's abuse of drugs. Although an individual may be addicted, if he or she does not know about dependence and withdrawal symptoms and if no one points these out as evidences of addiction (this is seen daily in millions of persons addicted to alcohol), the person may never actually admit to being an addict. Even someone with a mature personality needs strong social pressures or adult guidance to reverse the course of drug abuse after establishing a drug dependence. If drug abuse is continued, it will alter and possibly even destroy a person's life.

INTERACTIONS OF FACTORS

Although the initial drug use and selection of a specific drug or drugs is generally dictated by social circumstances (availability, friends, and social environment), individuals tend to obtain experiences from drugs related to their own basic needs. Psychological processes can override physical drug effects. Often people experiment until they find a drug whose physical properties produce the state of consciousness that fills their needs. For example, someone with a great need to escape from reality is likely to turn to a drug in the depressant or sedative group.

Emotional impairment becomes more important in drug abuse, and social factors become less important as use becomes heavier and more compulsive. The physical properties of different drugs fulfill specific needs created by corresponding impairments of ego function, and multiple-drug abuse compensates for a variety of such impairments. For instance, depressant drugs remove an individual from the stresses and anxieties of society. Heroin does this very quickly; alcohol, slowly but for a longer period of time. Hallucinogenic drugs may be used to compensate for a lack of experiences, whereas amphetamines may give a person an increased feeling of energy.

There is no single reason for a person to begin abusing drugs, no single pattern of abuse, and no inevitable outcome. Drug abusers are individuals and susceptible to many factors that are responsible not only for sustained drug use but also for recovery or for relapses. These factors are important not only in initiating the drug abuse and maintaining it but also in bringing about a drug cure. Just as drug abuse is motivated by a combination of social and individual factors, so is an individual's recovery from drug dependency usually brought about by a combination of factors.

SUMMARY

I. Patterns of drug abuse
  A. Drug behavior
    1. Experimenters are those who try drugs only a few times.
    2. Occasional users need the social or personal gratification drugs give them. This group includes "social drinkers" and "recreational drug users."
    3. Regular users take drugs one or more times a week and may be part of the drug subculture.

B. Drug dependency
   1. Habituation is psychological dependence. This condition may or may not include development of tolerance (need to increase dosage to attain same effect).
   2. Addiction is a state of physical as well as psychological dependency. In this state discontinuing of substance use leads to withdrawal illness.
C. Development of drug dependency
   1. Drug use continues if reinforced by personal or social reward.
   2. Continued drug use can be motivated by psychological, social, or physical factors.
D. New look at drug dependency
   1. Drug dependency has functions beyond the effects of the drug itself.
      a. It is a way of life.
      b. It is a way of dealing with uncomfortable feelings.
      c. It is a means of achieving in fantasy what cannot be attained in reality.
      d. It expresses anger and represses anger.
      e. It provides individual and group identity.
   2. People can let go of their substance dependencies when these needs are fulfilled in more satisfactory ways.

II. Psychological factors in drug abuse
   A. A "model" of emotional health and the role of drugs:
      1. Every person has basic needs.
      2. One element of emotional health is the development of effective and appropriate methods of fulfilling these needs.
      3. There is always some frustration of a person's needs.
      4. Thus, another element of emotional health is the ability to deal effectively with the frustration of your needs.
      5. The excessive use of alcohol or other drugs is an inappropriate method of trying to fulfill needs or of dealing with the feelings of frustration resulting from unfulfilled needs.
      6. The rehabilitation of many drug-dependent persons requires their development of adequate need-fulfillment techniques and/or improved methods of coping with need-frustration.
   B. Personality
      1. Personality is the total reaction of a person to his or her environment.
      2. The Freudian concept of personality includes:
         a. Id processes—inborn survival tendencies.
         b. Ego processes—the conscious mind, which develops gradually.

c. Superego processes—judgment mechanisms regarding right and wrong, good and evil.
C. Basic human needs (in sequence from most basic to higher needs) motivate much drug abuse:
1. Physiological needs.
2. Safety and security needs.
3. Need for love and belonging.
4. Esteem needs.
5. Need for self-actualization.
6. Need to know and understand.
7. Aesthetic needs.
D. Emotional dysorganization—There are five levels in order of increasing difficulty in coping, and each is related to drug abuse:
1. First level—commonly called "nervousness."
2. Second level—commonly called "neurosis."
3. Third level—escape of aggressive impulses from control of ego.
4. Fourth level—serious emotional disorders.
5. Fifth level—loss of will to live.

III. Sociological factors in drug abuse
A. Social factors may motivate drug abuse; drug abuse may have social effects.
B. Drug abuse occurs in all socioeconomic strata.
C. Abuse of specific drugs can often be associated with certain social or cultural groups within the population.
D. Signs of possible drug abuse in one's child:
1. Is secretive about friends and activities.
2. Expresses a negative attitude.
3. Is slipping in school.
4. Has drowsy spells during the day.
5. Experiences periods of undue excitement.
6. Has an excessive need of money.
7. Keeps arms covered.
8. Keeps a "stash."

IV. Interactions of factors
A. Drug abuse is motivated by a combination of social and individual factors.
B. Recovery from drug dependency is usually brought about by a combination of factors.

5

ENFORCEMENT AND/OR TREATMENT?

The earliest drugs to create abuse problems were the narcotics, which were being imported into the Americas even before the founding of the Republic. By the late 1800s the volume had increased until their use was common. The "miracle medicines," "elixirs," and "tonics," which contained large amounts of narcotics, usually tincture of opium (i.e., opium mixed with alcohol), were sold everywhere and were reputed to be cures for everything. Just before the Civil War, a further use of narcotics appeared with the invention of the hypodermic needle. Doctors encouraged their patients to buy this device and to use the "miracle" narcotics on a do-it-yourself basis.

By the end of the Civil War, thousands of soldiers had received large and numerous injections of narcotics to relieve their suffering from wounds and sickness. Many became highly addicted and relied completely on the drugs. With the growth of advertising and the promotion of patent medicines (containing large amounts of opium), great segments of the population used these medicines and became dependent upon them. Later, individuals found out about the specific narcotic ingredient contained in such medicines (usually opium or morphine) and started using it directly. Narcotic abuse climbed steeply until 1914 and stayed high even after the first drug-control laws were enacted. There was very little actual reduction in the use of narcotics by the general population until the Bureau of Narcotics was established in 1930 to enforce the earlier narcotics laws and apprehend violators. The name of this agency has since been changed to the Drug Enforcement Administration (DEA).

The major narcotic misused today is heroin, which was first introduced in 1898 as a cure for morphine addiction. In recent years other potent drugs have been produced that are used and abused to an even greater extent than the narcotics. For example, in the United States millions of people legitimately use amphetamines ("pep" and "diet" pills) and barbiturates (mainly sleeping pills). There are more mood-modifying drugs being developed today than any other group of drugs. The possibility of abuse is thus greatly increased by the familiarity of legally used drugs and the availability of new and more potent drugs.

PHILOSOPHIES OF DRUG-ABUSE CONTROL

There are at least two distinct philosophies concerning individuals who misuse drugs. One philosophy sees the drug offenders as people whose criminality manifests itself in drug offenses. Their involvement in the use, possession, sale, or theft of drugs is regarded as part of a predictable criminal behavior. The other philosophy views drug users as sick people who must be protected from themselves and prevented from spreading their disease to others. According to this philosophy, active addicts and drug users in the community represent a reservoir of infection, with drugs as the infecting agents and pushers as the vectors. From this idea, it follows that the smaller the reservoir (fewer number of drug users) and the scarcer the infecting agent (illicit supplies of drugs) and vectors (pushers), the less the danger of spread of the disease.

Most drug-control laws are directed toward the criminality of the individual, not the illness. These do not attack the actual drug problem but jail the offender. Since it is obvious that the misuse of drugs is a medical problem that has legal consequences, the abuse of drugs should be attacked as such. Those persons who become compulsive users and addicts should be studied for common traits in order to find the conditions, both social and psychological, that produce drug abuse. Then it may be possible to arrive at workable methods of treatment, rehabilitation, and control.

Actually, law enforcement is the means of control that must be used until research provides the knowledge that is needed to prevent drug abuse. Then law enforcement would be treated as a remedial action.

LEGAL CONTROLS OF DRUG ABUSE

Legal controls such as quarantines, isolation, and penalties have always been necessary to stop the spread of various diseases and illnesses. Since the best evidence supports the view that compulsive drug abuse is an indication of an emotional illness in an individual, society is justified in insisting on some type of regulation on the manufacture, distribution, and use of drugs. Such regulation can be viewed as part of preventive medicine.

Compulsive drug users or addicts are sick persons. They, however, seem quite willing to transmit the sickness to others. Therefore, to prevent the transmission of drugs, drug-abuse laws have been established. These laws are concerned with the illegal possession, manufacture, and sale of drugs, rather than with their use (which is a medical problem).

Federal Laws

The federal government's control of the sale and possession of narcotics and dangerous drugs is based on laws that have been enacted over the last 60 years. By tracing the history of these laws, we will be recapitulating the history of drug abuse in this country.

The first federal measure seeking control over drugs, enacted by Congress on February 9, 1906, was the *Federal Pure Food and Drugs Act*. It required food and drug manufacturers to list the amounts of alcohol and "habit-forming" drugs on the labels of their products. It did not restrict the sale or use of any of these materials. It also prohibited the importation or sale of opiates except for medicinal purposes.

A second law was enacted in 1914, the *Harrison Narcotics Act,* which further restricted the importation, manufacture, sale, and dispensing of "narcotics." It established the legal definition of a narcotic as *any drug that produces sleep and relieves pain*. Specific drugs were legally labeled narcotics regardless of their medical nature. Opium and its derivatives, coca leaves and their derivatives (such as cocaine), marijuana, peyote (mescaline), and any synthetic drug that produces sleep or stupor and relieves pain was declared a "habit-forming narcotic." The Harrison Act also required physicians to dispense narcotics only "in the course of their professional practice" and limited the selling of narcotics to licensed druggists and only when they received a lawful written prescription issued by a qualified medical or dental practitioner. It made the possession of narcotics without a prescription a criminal offense (felony).

The next federal statute, approved in 1922, an extensive revision of the Harrison Act, was the *Narcotic Drugs Import and Export Act*. It authorized the importation of specific quantities of crude opium and coca leaves needed to provide for the medical and scientific needs of the United States. It specifically prohibited the importation of opium for smoking or for the manufacture of heroin. This part of the law made it illegal to possess heroin in any form in the United States.

The fourth federal law, known as the *Marijuana Tax Act,* was patterned after the Harrison Act and was enacted in 1937. This act later became part of the Internal Revenue Code and is actually a tax law. The statute requires the registration and payment of a tax by all persons who import, manufacture, produce, compound, sell, deal in, dispense, prescribe, administer, or give away marijuana. At the time, the penalties were still the penalties for conviction under a narcotics law, and marijuana was still classified as a narcotic.

The control of other drugs has not been consistent with the corresponding controls of narcotics. In 1938 the *Federal Food, Drug and Cosmetic Act* was enacted. This law divided drugs into two main classes: prescription drugs and over-the-counter drugs. Many drugs are safe and effective if used according to directions on the label. Such drugs, according to this law, may be sold without a prescription. But when a drug is so powerful that it should be used only under a physician's supervision, the law requires that it be sold only by prescription. In essence, this is the basis of the American public's "self-medication" abilities. The penalties under this law consisted of fines and misdemeanor charges against offenders.

In 1942 the *Opium Poppy Control Act* was approved. This act was passed when World War II cut off the supply of opium from Asia to the United States. It outlined the requirements for obtaining a license from the federal government for the cultivation of opium poppies in the United States if that should ever be needed. The development of synthetic narcotics has minimized the likelihood that a scarcity will ever occur.

Until 1951 the control of narcotics was directed toward the medical profession and their dispensing of these drugs. In 1951 the *Boggs Act* was passed, spelling out the penalties to be given to those convicted of violating the federal narcotic laws. It stipulated that anyone who violated the federal laws may

> be fined not more than $2,000 and imprisoned not less than two or more than five years. For a second offense, the offender shall be fined not more than $2,000 and imprisoned not less than five or more than ten years. For a third or subsequent offense, the offender shall be fined not more than $2,000 and imprisoned not less than ten or more than twenty years.

This mandatory-sentence law provided severe penalties and limited the suspension of sentences or the granting of probation or parole to a first offender only.

Also in 1951 the prescription requirements of the 1938 Federal Food, Drug and Cosmetic Act, along with controls on distribution, were strengthened by the *Durham–Humphrey Act*. This amendment was directed against barbiturates and amphetamines, making it illegal to dispense them without a prescription. It still retained the misdemeanor penalties of the 1938 law.

The penalties for violations of the federal narcotics laws became even heavier in 1956, when the *Narcotic Control Act* was passed. This law increased sentences at all levels until a third offender was given a mandatory 10–40 years with no possibility of probation, suspension of sentence, or parole.

But the major point of the Narcotic Control Act is that it also provided that

> any person having attained the age of eighteen years, who knowingly sells, gives away, furnishes, dispenses, or conspires to sell, give away, furnish, or dispense any heroin to any person who has not attained the age of eighteen years, may be fined not more than $20,000, imprisoned for life, or suffer death if the jury, in its discretion, shall so direct.

The Drug Enforcement Administration maintains that these severe penalties act as a powerful deterrent. Medical groups do not agree. They do not believe that narcotic addicts who are physically and psychologically dependent on a drug will forego satisfaction of this craving for fear of a long prison sentence or that they will be deterred by fear of the penalty that awaits them if they are caught possessing it. Both positions are valid, but the deterrent position would be much stronger if the potential victim understood the penalties prior to their first introduction to narcotics.

The need for more stringent controls over the manufacture, distribution, and illegal abuse of mood-modifying drugs other than the illegal narcotics (opiates, cocaine, and marijuana) became a point of focus in the early 1960s. Thus, in 1965 the *Drug Abuse Control Amendments* to the 1938 Federal Food, Drug and Cosmetic Act were enacted by Congress. This law established the legal definition of a "dangerous drug" as barbiturates, amphetamines, LSD, and any other type of drug except narcotics (opiates, synthetic narcotics, cocaine, or marijuana) that has been shown to have a potential for abuse because of its depressant, stimulant, or hallucinogenic effects on man, causing mood modifications or abnormal social behavior. These drugs were considered "controlled dangerous drugs," which set them apart from the "narcotics" of the Harrison Act. The penalties sections of the 1965 amendments were further amended in 1968.

Under the provisions of this federal law as amended, all wholesalers, jobbers, and manufacturers of controlled dangerous drugs must register annually with the Food and Drug Administration and keep records of sales of these drugs. Pharmacists, hospitals, researchers, and physicians who regularly dispense and charge for the controlled drugs must maintain records that are available for inspection by the Food and Drug Administration. Prohibitions include refilling a prescription for any one of the controlled drugs more than five times or later than six months after the prescription was originally written; manufacturing, processing, and compounding the controlled drugs, except by registered drug firms; and distributing the designated controlled drugs to persons not authorized by federal or state laws to receive them. Illegal possession of dangerous drugs was made punishable by a misdemeanor penalty. Strong felony penalties were provided for those who illegally produce, sell, or dispose of dangerous drugs. The Food and Drug Administration inspectors were given more authority and stronger enforcement powers to prevent drug counterfeiting and other violations of the law.

Even with the passage of extreme penalties for violation of the federal narcotics laws in 1956 and of the stronger controls of dangerous drugs in 1965, the abuse of drugs continued upward. By 1966 the opinions of medical authorities were being heard and a trend toward treatment and rehabilitation of drug users started to develop. The first step toward this was passage of the *Narcotic Addict Rehabilitation Act* of 1966. This act changed the status of the two federal narcotics hospitals (one at Fort Worth, Texas, and the other at Lexington, Kentucky) from treatment centers to research centers. The act also allocated money to states and local communities for the treatment of drug abusers.

In 1970 some semblance of consistency in the federal drug control laws was established with the *Comprehensive Drug Abuse Prevention and Control Act*. This was a new schedule of federal drug penalties intended to replace and update older drug laws to fit the needs of the drug-abuse problems of the 1970s.

The main provision of this law established five classes of drugs whose illegal manufacture, distribution, possession for use, possession for sale, and sale are controlled. An outline of the five classes and the penalties for violations are shown in Table 5.1.

Class 1 drugs are considered to have the highest potential for being abused because of their mood-modifying qualities. They carry the most severe penalties, are regarded as the most dangerous, and are outlawed even for medical use in the United States. Class 2 drugs are medically used drugs that have the same potential for abuse as Class 1 drugs and that carry the same penalty structure as Class 1 drugs. Class 3 drugs are considered to have a potential for abuse but not as high a potential as Class 1 and 2 drugs. Class 4 drugs have a lower potential for abuse; their penalties are further reduced. Class 5 drugs have the lowest potential for abuse, and the penalties are the mildest.

The law also gives the U.S. attorney general the power to decide to which class a new drug belongs on the basis of its "potential for abuse." The attorney general may also move a drug up or down the schedule. Before doing this, a scientific panel appointed by the president must be consulted.

With the new law, marijuana has been reclassified as a dangerous drug and is handled under the nonnarcotic penalty structure as a Class 1 drug; individual possession of marijuana is now a misdemeanor. Minimum mandatory penalties for possession of any drug were eliminated. And a maximum penalty of one year for possession (first offense) and three years for subsequent offenses was established. Also, for a first offense, an individual under 21 years of age who is convicted of possession may be placed on probation (without sentencing); if probation is successfully completed, the official arrest, trial, and conviction can be erased from the record.

State and Local Laws

In drug control there has always been cooperation between federal, state, and local law enforcement officers. Until a few years ago, the problem of enforcement was left largely to the former federal Bureau of Narcotics. Now there are separate narcotics units in local police departments of all large cities, and a large proportion of drug-abuse cases are prosecuted in local and state courts.

In 1932 a model uniform state narcotics law patterned after the Harrison Narcotics Act was submitted to several state legislatures. Since that time, this law, with minor changes in some states, has been enacted by 47 state legislatures, by Puerto Rico, and by Congress for the District of Columbia. California and Pennsylvania have enacted laws that appear to be comparable in scope and effectiveness. Some states, such as Ohio and Minnesota, have enacted even heavier penalties. For instance, in 1955 Ohio enacted legislation providing for a 20-year minimum penalty for the unlawful sale of narcotics. And in many other states penalties range from 2–10 years of imprisonment for a first offense of unlawful possession to 5–20 years for the unlawful sale of narcotics.

However, state laws dealing with the dangerous drugs are anything but uniform. All states have some type of regulation covering prescription drugs and many laws specifically regulating barbiturates and amphetamines. The current practice of the states in dealing with the hallucinogenic drugs is to remodel old laws or adopt new ones that conform more closely to the Drug Abuse Control Amendments. Since most states have no clinics or laboratories in which to evaluate possible dangerous drugs or narcotics, they depend upon the recommendations made by the federal government. States usually enact laws against the abuse of those drugs that federal laws have declared to be dangerous drugs or narcotics. Nevada and California reversed this procedure by enacting laws governing LSD prior to the enactment of federal regulations.

Many states are currently in the process of examining their drug laws, and some are making extensive changes. Prosecution to the limit of the laws is rare. Also, judges have great flexibility in interpreting and applying these laws. Since the passage of the federal Narcotic Addict Rehabilitation Act of 1966, many states have shifted their concern more toward laws concerning rehabilitation rather than punishment of drug abusers.

There are other trends in state law enforcement. In New York State, the disposition of an individual convicted of a drug-abuse crime is dependent upon the degree of the offense. For example, if an individual is convicted of possessing fewer than 25 marijuana cigarettes, the crime is

TABLE 5.1
Schedules and Penalties for Violation of the Comprehensive Drug Abuse Prevention and Control Act of 1970[a]

Drug Schedule	Potential for Abuse	Medical Use	Production Controls	Examples of Drugs in Each Class	Maximum Penalties for Manufacturing and Distribution	Maximum Penalties for Simple Illegal Possession
Class 1	High	None	Yes	Opium derivatives and hallucinogens, such as heroin, marijuana, THC, LSD, and mescaline	*Narcotics* First offense: 4–15 years; $15,000 fine Second and subsequent offenses: 6–30 years; $50,000 fine	First offense: up to 1 year, (probation possible); $5,000 fine Second offense: up to 2 years; $10,000 fine
Class 2	High	Yes	Yes	Medically utilized narcotics and injectable metamphetamines, such as morphine, cocaine, and methadone	*Nonnarcotics* First offense, 2–5 years; $15,000 fine Second and subsequent offenses: 4–10 years; 30,000 fine	
Class 3	Moderately high	Yes	None	Mild narcotics such as codeine, amphetamines, and barbiturates	First offense: 2–5 years; $15,000 fine Second and subsequent offenses: 4–10 years; $20,000 fine	
Class 4	Low	Yes	None	Mild barbiturates such as chloral hydrate and meprobamate	First offense: 1–3 years; $10,000 fine Second and subsequent offenses: 2–6 years; $20,000 fine	
Class 5	Quite low	Yes	None	Low percentage narcotic mixtures, such as tranquilizers	First offense: up to 1 year; $5,000 fine Second and subsequent offenses: up to 2 years; $10,000 fine	

* Schedules and penalties may be changed by the U.S. attorney general at any time.

considered a misdemeanor; possession of over 25 marijuana cigarettes is considered a felony. Also in keeping with this trend of flexibility in law enforcement is legislation passed in California in 1968 allowing a judge to sentence an individual convicted of possession of marijuana (first offense) to one year or less in county jail (a misdemeanor sentence) or to a state prison for 1–10 years (a felony sentence).

As an example of specific state drug-abuse-control laws, the drug-abuse laws of the state of California reinforce those of the federal government and in some cases even exceed them in severity. According to these laws, the illegal possession, sale, or administration of any narcotic drug is a felony while visiting a place where they are being sold is a misdemeanor. Also, in California it is a felony to cultivate or process marijuana, while the possession of paraphernalia for the use of marijuana is not illegal.

In 1972 California established five classes for narcotics and "restricted dangerous drugs" (see Table 5.1). This brought California's drug laws into closer conformity with the federal Drug Abuse and Control Act of 1970. LSD is restricted, and its use for purposes other than authorized research is specifically prohibited by California law. This law is a boon for scientific research in California because although prohibiting casual use, the law allows research into the effects of the drug. Thus, the specific problems associated with the abuse of this drug may someday be better understood.

The triplicate-prescription provisions of the California Narcotic Act provide a control over possible abuses relating to medically prescribed narcotic drugs. According to the provisions, physicians must write prescriptions for narcotics in triplicate on official blanks (Figure 5.1). One copy of each prescription is filed with the California State Bureau of Narcotic Enforcement; one copy is kept on file in the physician's office; and one is held by the pharmacist who dispenses the drug. Thus, excessive prescription of drugs by a physician, excessive use by a patient, or irregularities in the handling of narcotics by a pharmacist can be quickly investigated by the Bureau of Narcotic Enforcement. Even more significant is the effectiveness of the triplicate-prescription procedure in enabling the bureau to detect the diversion of legitimate drug supplies into illicit channels, as well as enabling it to uncover the fraudulent writing of prescriptions by unauthorized persons.

Not all narcotic preparations are treated alike in all states. Some narcotic-containing medicines, classified under the law as "exempt" preparations, can be sold without a physician's prescription. The narcotic content of these medicines (usually cough syrups) is very low. At present some states limit the amount sold to one person within a 24-to-48-hour period. In a few states a person must be over 21 years of age in order to buy these drug products without a physician's prescription.

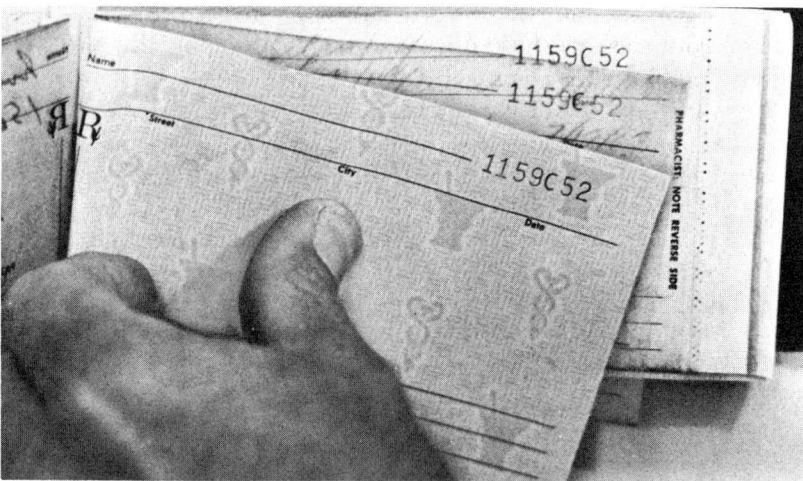

Figure 5.1. Triplicate narcotics prescription form used in California. Narcotics prescriptions are written in triplicate and then bound as a book so that when the patient is given the original prescription and a duplicate, a tissue copy is retained in the book as a permanent record for the physician. The two copies given to the patient are taken by the patient to a pharmacy. There, after filling the prescription, the pharmacist endorses the duplicate, which is sent to the Bureau of Narcotic Enforcement at the end of each month. The original remains in the pharmacist's files as a permanent record.

Other states, such as California, require a prescription for all narcotic-containing preparations.

DRUG CONTROL AND LAW ENFORCEMENT

Control of drug use in the United States has had a mixed and confused history. Prohibition of alcoholic beverages during the 1920s is generally considered to have been a failure. Yet, control-by-prescription laws have certainly determined that drugs will not be routinely abused. A person cannot simply walk into a drugstore and purchase morphine without a prescription; a five-year-old child cannot purchase even aspirin by himself. Surely the legal requirements that certain substances be dispensed only upon the approval and under the direction of a physician have affected most people's attitudes toward drugs.

One of the most interesting aspects of drug control is the current argument centering on the possible legalization of marijuana. Proponents of legalization have insisted that prohibition of alcohol during the 1920s did not work and that the restrictions on the cultivation, sale, use, and possession of marijuana are not working either.

Proponents of marijuana indicate that although current research shows that marijuana does little to the body that alcohol does not do, society has hypocritically banned one and permitted the other. The major conclusion to be drawn from this argument on legalization of marijuana is that American society has the most difficulty in developing useful policies toward those substances that occupy a middle ground in the public consciousness. Alcohol, tobacco, and marijuana have attained such an undefined position primarily because their effects are not readily predictable and because their use does not necessarily interfere with normal functioning in society.

Laws are functional representations of what a society believes in. A society has laws against forms of conduct it condemns and laws designed to encourage conduct of which it approves. Along with other factors, the body of laws serves as an instrument of continuity between generations, providing new and young members of the society with guidelines for their behavior and attitudes.

Helen H. Nowlis, professor of psychology at the University of Rochester, has outlined nine points that help explain how attitudes toward the problem of drug abuse develop.

1. *Ignorance* of the actions of chemical substances upon the complex chemical system that is the living person can be a large factor. There is also a lack of knowledge about the relationship of variations in human behavior to human behavior itself. In the absence of knowledge, drug abuse is a problem of opinion, emotions, and beliefs.

2. *Semantics* are important because every term can be entangled in myth, emotions, assumptions, beliefs, and attitudes that too often turn the dialogue into a futile argument.
3. *Communication* between scientists themselves and with the public has been a significant problem in American society for decades. Technical jargon and scientific precepts and assumptions are frequently very difficult for the layperson to comprehend. There is also a communication problem between a generation brought up before and during the development of automation, television, jet travel, nuclear energy, large urban centers, cramped schools, and an affluent social order and a generation that knows no other way of life.
4. *Lack of understanding of scientific method and concepts* contributes to the problem because few people understand that there are seldom simple cause-and-effect relationships in dealing with human behavior. The term "proof," which people seek in determining their attitudes, cannot always be used reliably. The design and execution of experiments might be open to bias; conclusions might be colored by the scientist's own viewpoint.
5. *Living in a world of constant and dramatic change* has contributed to a disorientation of the traditional beliefs and assumptions of society. The future is unpredictable, and people have become increasingly concerned with the here and now.
6. *The philosophy of social control* is an increasingly difficult aspect of the problem. What should society prevent a person from doing to himself? How does the entire social structure suffer when an unjust or unenforceable law continues to diminish public respect for the judicial and legislative processes?
7. *Education* is both a major key to and an important part of the difficulties of determining social attitudes toward drugs. Educational institutions and the information media have an obligation to inform citizens of new developments in scientific research, but institutions of learning then become vulnerable to shortsighted attacks when people disagree with these conclusions or the institutions that produced them.
8. *American society is a pill society.* This well-advertised proposition states that there is a chemical solution for any problem of unpleasantness, strain, or discomfort. This pill society spends more on alcohol, tranquilizers, sleeping pills, and tobacco than it does on education or social and economic problems.
9. *Increasing retreat in the face of complex problems* is the end point of the process. As society becomes more complex, its confused and uncertain people seek more absolutes and find themselves less and less able to relate to the society and to understand their own roles. For many people, black-and-white choices seem easier than careful, deliberate decisions based on information and sound judgment.

It is becoming increasingly evident that people who have social, personal, and intellectual problems and who are unable to solve those problems abuse drugs. Individuals do not continue to do something that does not provide them with satisfaction. The reasons for abusing drugs are complex and exist whether the drugs are socially acceptable or not. These reasons center on desires to change the pace of living, to modify the demands society makes on the individual, to relieve boredom and distress.

As we have described, the major laws have until recently been established to control illegal possession, manufacture, and sale of drugs, rather than to find ways to help people cope with modern society and with their abuse of drugs. A trend is developing to establish laws that will govern research into the effects and actions of drugs and that will permit greater flexibility in the treatment of drug abuse and the rehabilitation of drug abusers.

DRUG-ABUSE TREATMENT

Some type of control is needed for recovery from drug abuse. This may be legal, social, or self-control; but it must be present. The ultimate success of such control will depend upon the motivation of the drug abuser to modify his or her behavior. The greater the evidence or promise of self-control, the lesser the need for extended legal restraints. The hope is for sufficient rehabilitation for the person to resume an independent, productive existence without further reliance upon illicit drugs. Where such an expectation is not realized, extended legal control may be necessary.

Each individual treatment program must be based upon the particular characteristics of the case. Volunteer programs provide the individual with a highly structured environment, the most important element of which is family or social control through peer populations. Individuals who lack even the self-control to stay within these frameworks and who leave a program against medical advice, will always be the ones who are placed within the legal controls of public institutions.

Treatment Laws

The federal Narcotic Addict Rehabilitation Act of 1966 was the first big step toward establishing public treatment programs and facilities. This act and other legislation passed from 1968 to 1971 gave federal support, through the National Institute of Mental Health, to develop training programs and to construct, staff, and operate addiction-treatment facilities within the individual states. This support has helped the states expand their programs and facilities to meet their increasing needs. Many states have also passed legislation supplementing and defining the procedures of these laws. In California there is the Mental Health Act of 1969 (the Lanterman-Petris–Short Act), which defines in detail the conditions that permit civil and criminal commitment for drug abuse. Figure 5.2 outlines the procedures for committing someone for treatment in the state of California.

Civil commitment for drug abuse. Civil commitment is a legal mechanism utilized to ensure control over drug abusers during rehabilitation, first in an institution, later in a halfway house, and still later in the community under the close supervision of a probation or parole officer. The significant point in this procedure is that the drug abuser does not establish a criminal record when seeking help. An individual is limited to 2½ years of maximum commitment unless the judge renews the commitment. If, after release to the community, they resume using drugs, they may be returned for further treatment.

If, at the time of civil commitment, the drug abuser was on trial because of possession, sale, or another drug-related offense, he or she may be returned to the court after completing the treatment program for dismissal of the charges or sentencing. Should he or she be found guilty of the criminal charges, any time served in the treatment program will be credited toward the time to be served under the criminal sentence.

Also, under this procedure, a physician, police officer, or relative of someone who is an addict can volunteer the addict for commitment by filing a petition with the federal district attorney, who may then order the alleged addict to appear for examination. At commitment, the process stops being voluntary. The term of treatment cannot exceed 2½ years.

Criminal commitment for drug abuse. If a judge believes an eligible convicted criminal offender is an addict, the abuser may be ordered to be examined and then committed to a treatment program for an indeterminate period of time. This cannot be more than seven years but must be more than six months. The offender then may be released under the supervision of a probation or parole officer.

All states and the federal government have procedures similar to these. Information may be obtained by telephoning any district attorney, police station, or sheriff's office; they will outline the procedures. Remember, drug abuse is not a crime, but possession of drugs is a crime. If the individual seeking treatment is "clean," there is no need to worry about proceedings.

Drug-Abuse Therapy

Not everyone who explores the effects of drugs immediately becomes a compulsive user. Some use drugs only occasionally; others may use them

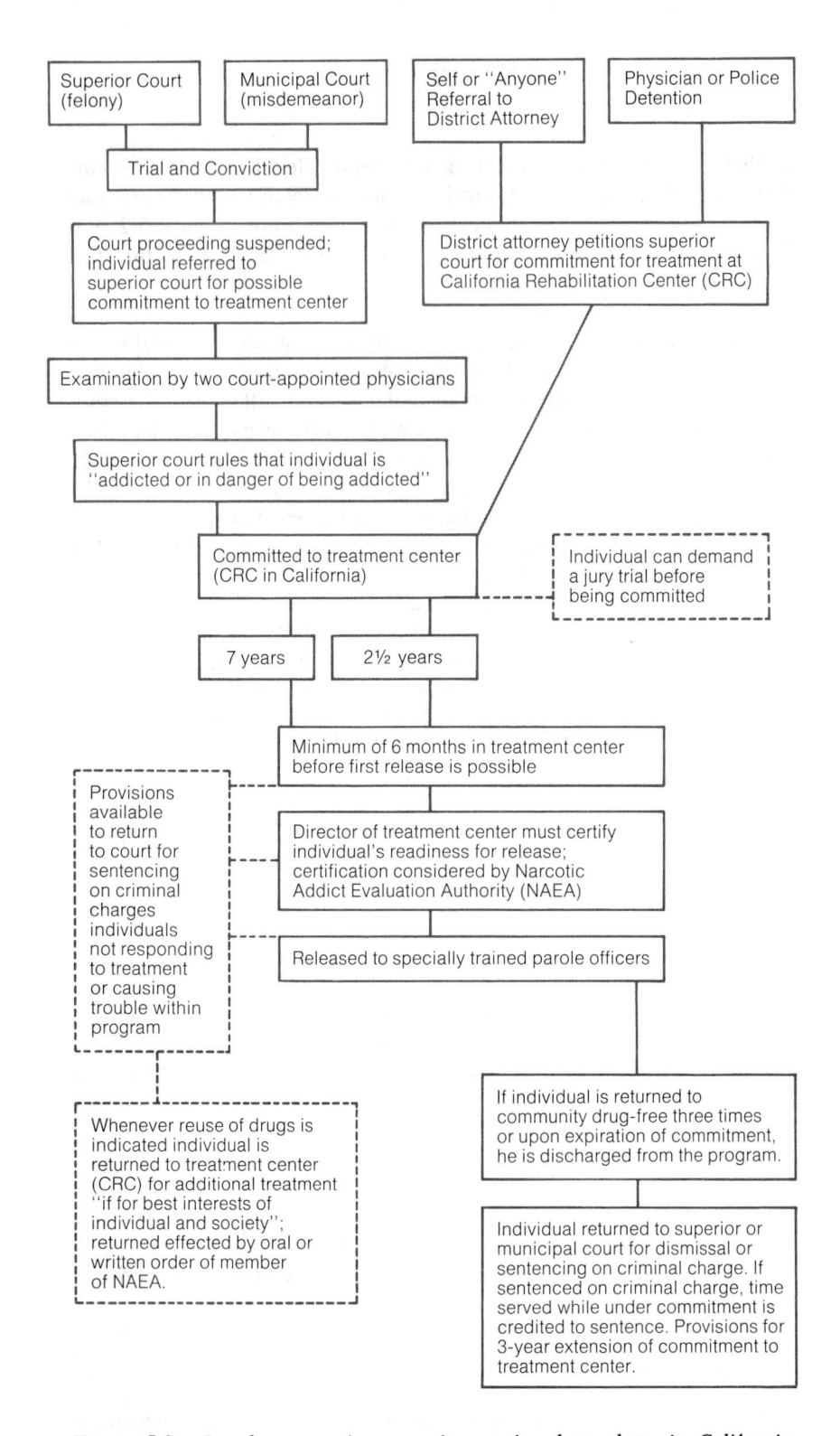

```
┌─────────────────┐ ┌─────────────────┐   ┌─────────────────┐ ┌─────────────────┐
│ Superior Court  │ │ Municipal Court │   │ Self or "Anyone"│ │ Physician or    │
│ (felony)        │ │ (misdemeanor)   │   │ Referral to     │ │ Police          │
│                 │ │                 │   │ District Attorney│ │ Detention      │
└────────┬────────┘ └────────┬────────┘   └────────┬────────┘ └────────┬────────┘
         └──────────┬────────┘                     └──────────┬────────┘
         ┌──────────┴──────────┐                              │
         │ Trial and Conviction│                              │
         └──────────┬──────────┘                              │
         ┌──────────┴──────────────────┐    ┌─────────────────┴─────────────────┐
         │ Court proceeding suspended; │    │ District attorney petitions superior│
         │ individual referred to      │    │ court for commitment for treatment at│
         │ superior court for possible │    │ California Rehabilitation Center (CRC)│
         │ commitment to treatment center│  └─────────────────┬─────────────────┘
         └──────────┬──────────────────┘                      │
         ┌──────────┴────────────────────────┐                │
         │ Examination by two court-appointed │               │
         │ physicians                         │               │
         └──────────┬────────────────────────┘                │
         ┌──────────┴──────────────────┐                      │
         │ Superior court rules that   │                      │
         │ individual is "addicted or  │                      │
         │ in danger of being addicted"│                      │
         └──────────┬──────────────────┘                      │
                    │                                         │
              ┌─────┴─────────────────────────────────────────┘
              │
    ┌─────────┴─────────────────┐         ┌─────────────────────────────┐
    │ Committed to treatment    │ - - - - │ Individual can demand       │
    │ center (CRC in California)│         │ a jury trial before         │
    └─────────┬─────────────────┘         │ being committed             │
              │                           └─────────────────────────────┘
         ┌────┴────┐
    ┌────┴───┐ ┌───┴─────┐
    │7 years │ │2½ years │
    └────┬───┘ └───┬─────┘
         └────┬────┘
    ┌────────┴─────────────────────────────┐
    │ Minimum of 6 months in treatment     │
    │ center before first release is possible│
    └────────┬─────────────────────────────┘
┌──────────────┐ ┌────────┴─────────────────────────────┐
│ Provisions   │ │ Director of treatment center must    │
│ available    │ │ certify individual's readiness for   │
│ to return    │-│ release; certification considered by │
│ to court for │ │ Narcotic Addict Evaluation Authority │
│ sentencing   │ │ (NAEA)                               │
│ on criminal  │ └────────┬─────────────────────────────┘
│ charges      │ ┌────────┴─────────────────────────────┐
│ individuals  │ │ Released to specially trained parole │
│ not responding│ │ officers                            │
│ to treatment │ └────────┬─────────────────────────────┘
│ or causing   │          │
│ trouble within│         │      ┌─────────────────────────────────┐
│ program      │          ├──────│ If individual is returned to    │
└──────┬───────┘          │      │ community drug-free three times │
       │                  │      │ or upon expiration of commitment,│
       │                  │      │ he is discharged from the program.│
       │                  │      └─────────────────────────────────┘
┌──────┴────────────────────────────┐ ┌─────────────────────────────────┐
│ Whenever reuse of drugs is        │ │ Individual returned to superior │
│ indicated individual is           │ │ or municipal court for dismissal│
│ returned to treatment center      │ │ or sentencing on criminal charge.│
│ (CRC) for additional treatment    │ │ If sentenced on criminal charge,│
│ "if for best interests of         │ │ time served while under         │
│ individual and society";          │ │ commitment is credited to       │
│ returned effected by oral or      │ │ sentence. Provisions for        │
│ written order of member           │ │ 3-year extension of commitment  │
│ of NAEA.                          │ │ to treatment center.            │
└───────────────────────────────────┘ └─────────────────────────────────┘
```

Figure 5.2. Legal process for commitment for drug abuse in California.

intermittently; and others may progress over many months to regular and constant use. There is no sharp line that divides the regular user from the constant, compulsive user, or addict. Such abuse is only an extreme on a *behavioral continuum*. The treatment of drug abuse should never be directed toward the drug. It must be aimed at the problems of individuals abusing drugs. Then, to a great extent, it will be possible to limit drugs to appropriate use if people can be provided with a purpose in life that is more important than drugs. There must also be recognition of the various patterns of drug use in order to avoid handicapping a person with the undeserved label of "drug addict."

There are no proven techniques or procedures that can be applied to the total drug-abusing population. The methods of drug-abuse treatment currently used at institutions in the United States are highly varied in their approaches and results. Yet, irrespective of these facts, all drug-abuse treatment programs ultimately depend on the motivation of the individual. By placing final responsibility and hope for cure on the will of the individual, society has also implied that the alteration of certain external factors will not be sufficient to reverse the tragic patterns of abuse that have occurred in recent decades.

Modes of Therapy

Medical authorities feel that the compulsive drug abuser is a sick person, emotionally disturbed and often physiologically ill. Treatment is needed for the physical effects of the drug being abused and psychological help is needed to keep the individual from going back to drug abuse after leaving the hospital.

Tests for drugs. One of the first tests for drug use was the Nalline test. It was used as early as 1955 by Dr. James Terry in Alameda, California. The Nalline test made possible for the first time the examination of large numbers of individuals for evidence of use of drugs. In this test Nalline (nalorphine) or Levallorphan is injected subcutaneously. If an opiate drug has been used within the last week, the individual will show (according to how much drug has been used and how recently) drastic constriction of the pupils, sweating, vomiting, or full withdrawal symptoms.

In 1970 the U.S. armed forces initiated a drug-testing program. It consists of a series of chemical assay tests on urine. The free radical assay technique (FRAT) test measures free and bound morphine (the end product of heroin metabolism in the body) in the urine. It is a rapid test and gives an answer one minute after the urine sample is mixed with the reagents and put into the machine. The thin layer chromatography (TLC) test measures barbiturates, amphetamines, opiates, methadone, and many other drugs, depending upon the reagents used. This test takes only four hours and will detect almost any drug the physician wants it to detect. If a positive reaction to either of these tests is shown, the individual is given the gas liquid chromatography (GLC) test which is 99 percent accurate in detecting any drug abuse by an individual. As equipment becomes available, the chemical tests will replace all other tests.

Initial emergency therapy. When an individual under the influence of drugs comes to the attention of a hospital staff or a private physician, it is usually in the context of an emergency situation. Some physicians in high-abuse areas will see as many as 7000 drug users in one year.

The first information needed by the physician is the name of the drug (or *combination* of drugs) the individual is using. This is often difficult to determine because the user may be unconscious, semicoherent, disoriented, frightened, or unreliable and may often behave like an acute psychotic. A reassuring approach in a quiet environment will often produce an accurate story of what happened and thus simplify the emergency phase of treatment. Friends, the contents of the user's pockets, and the surrounding conditions under which the adverse reactions occurred might help the physician's evaluation. During this acute phase of treatment, a well-established rapport between the physician and the patient can be a valuable tool. This is the talk-down phase. Little more than acceptance and reassurance is necessary. If definite signs of toxic complications or withdrawal occur, the physician will treat them as they appear, regardless of what the patient may have said. In some states the attending physician is prohibited by law from prescribing for, administering, or dispensing to an addict or habitual user any controlled drug.

Most drug prosecutions are based on evidence of possession. If the police obtained such evidence prior to bringing the patient to the health facility, such evidence is valid. If such evidence has not yet been obtained at the time of medical examination, the police will not continue into the examining room in their search. A physician is not a law enforcement officer but rather a person whose responsibility it is to help the patient through an extremely disturbing and dangerous experience. The physician needs the patient's greatest possible cooperation in order to treat successfully, and it is in the patient's best interests to understand this.

Withdrawal therapy. When treating a victim of addicting drugs, a physician has an even more complicated task. Abrupt withdrawal, the so-called "cold turkey" (as in the case of barbiturates), is painful and can be fatal. After the emergency is over, gradual withdrawal should take place in a hospital. The American Medical Association suggests that a physician should not normally attempt withdrawal unless the individual has been hospitalized.

When drugs such as opiates or barbiturates are used several times a day, the user becomes tolerant to and physically dependent upon the drug. Some cells and body systems become tolerant rapidly and others slowly. When a drug is withdrawn, the tolerant cells and body systems must return to normal. The symptoms produced while these cells are returning to normal are called withdrawal, or the abstinence syndrome.

During narcotics withdrawal, the addict experiences the abstinence syndrome, which includes nausea, watering of the eyes, muscle spasms in the stomach and legs, and hot and cold flashes. The synthetic narcotic *methadone* may be substituted for the addicting drug because methadone tends to block the euphoric effects and relieve the craving for drugs such as morphine and heroin.

In barbiturate withdrawal, convulsions similar to those in grand mal epilepsy and delirium tremens may occur. In this case the drug pentobarbital can be used as a substitute drug during withdrawal. Abrupt withdrawal from barbiturates is extremely dangerous and has been fatal. Consequently, a physician must know what drug an individual has been using before attempting withdrawal.

After an emergency situation and withdrawal (if necessary), many drug users (including experimenters, occasional users, and intermittent users), if given adequate supervision, may give up the use of illicit drugs after an enforced period of drug-free living. This if only to avoid an involuntary return to an institution. Of course it is unrealistic to expect every user to reach all the goals of ideal treatment. But the first realistic step by the abuser toward recovery is to be able to perform normally within society, even under supervision. There are a number of treatment methods available to help an individual achieve this.

Narcotic-antagonist therapy. One such treatment method has involved the temporary use of narcotic-resembling antagonists. *Narcotic antagonists* are drugs chemically and structurally so similar to narcotics that they can apparently occupy the receptor sites in the nervous system normally occupied by the illicit opiate. The difference is that the antagonist does not provide the abuser with the narcotic effect. When given to a person physically addicted to narcotics, the antagonist will bring on withdrawal systems, rather than prevent them (as do the illicit opiates). In this way, a physician can tell if an abuser is currently using narcotics.

Three of the most important antagonists are *Nalline, Cyclazocine,* and *Naloxone.* The antagonist effect of nalline is the basis for the once-widely-used Nalline test.

When taking daily dosages of an antagonist, such as cyclazocine, the patient will not feel the effects of a narcotic nor will he or she become addicted to it. Thus, during a treatment program, the narcotic antagonist may provide the addict with the opportunity for modifying his or her drug-related behavior. The antagonist gives the addict time in which to begin participating in a constructive rehabilitation program without the danger of becoming addicted to the antagonist. For full effectiveness, the use of antagonists *must* be accompanied by a comprehensive program of psychological and social rehabilitation.

Drug maintenance therapy. Because of the short-acting ups and downs of mood-modifying (psychotropic) drugs (especially heroin), a drug user is preoccupied with the thought of obtaining more drugs. In order to establish the first step in treatment—return to normal social functioning—medical therapy may be needed to eliminate the need for psychotropic drugs and at the same time discourage their use. The drug *methadone* and a few others, such as *acetylmethadol,* can be substituted for any major opiate drug, block the euphoric effects, and relieve the craving for other opiates.

In 1965, Dr. Vincent Dole and Dr. Marie Nyswander, both of Rockefeller University in New York City, developed the first systematic methadone-maintenance program. In this program the patient is given a quantity of methadone on a daily basis and thus has the opportunity to live a fairly normal life. Unlike the narcotics addict who is relying solely on illegal opiates, the individual being treated with methadone need not focus all efforts upon obtaining further illicit drugs. Methadone thus permits the individual a more immediate, responsible view of self and society. This program has now spread to all states and is a major form of treatment. Many individuals have become acceptable citizens through a treatment program that combines methadone maintenance (blockage) with psychological and social rehabilitation.

Drugs such as methadone, pentobarbital, and acetylmethadol have certainly made possible some degree of progress in treating serious drug-abuse cases. But these methods have also drawn a good deal of criticism. One reason for this criticism is the observation that these procedures succeed only with highly motivated individuals. Also, many physicians feel that these studies are based on the ethically unacceptable scheme of substituting one addictive drug (methadone) for another (heroin). The New York Academy of Medicine concedes that methadone maintenance is not the ultimate cure for heroin addiction, but it points out that "no other regimen currently available offers so much to the chronic addict."

Therapeutic communities. Another distinct approach to drug-abuse therapy involves the establishment of complex social systems, houses, or communities that are directed almost exclusively by ex-addicts. Such organizations as Synanon (California-based), Daytop Village, Phoenix House, Odyssey House (all in New York), and Gateway House (Illinois) are some of the better-known therapeutic communities. Individuals are not required to remain at these centers and are free to leave permanently at any time. However, in order to remain at the center, they must participate in the center's programs and must conform to the strict community rules.

Therapeutic communities are designed to provide care through three mechanisms: (1) encounter-group therapy, (2) a highly structured community organization, and (3) a reward–punishment system based on simple behavioral psychology. The key to the therapeutic process is the group encounter, usually called the "Synanon game," "attack therapy," or the "verbal street fight." The second phase consists of a complete behavioral dissection through the encounter therapy and a scaled program of house jobs, starting with dishwashing or garbage control, and progressing to ordering supplies and leading group therapy sessions (this phase can last for years).

These communities are designed to develop and reinforce the standard middle-class norms of behavior and attitudes. Theoretically, the community tries to make the individual goal-oriented: willing to work hard and sacrifice in order to obtain eventual security and success. Men's and women's roles are carefully defined, and sexual behavior is subject to the censure and influences the individual would encounter outside the community (homosexual relationships are prohibited and heterosexual relationships are permitted after an orientation period in the community).

Finally, reentry into society is the hoped-for expectation of the program. The addict is progressively exposed to the world outside the community. First, a job may be taken inside the community. Next, the sphere of activities will move further outside the boundaries of the community, including a job and residence on the outside.

Some therapeutic communities are criticized for the inability of their residents to enter the community. Some, such as Synanon, no longer encourage any addict to leave the community and reenter society. Their experience indicates that the addict usually requires the permanent support of the therapeutic community in order to prevent the resumption of drug-taking behavior.

Very few individuals in therapeutic communities ever achieve full-time employment in positions unrelated to addiction-treatment programs. Thus therapeutic communities may fail to make people independent, yet provide a life that is better than that of the addict. Some view them as encouraging addiction to the community life rather than to hard drugs. Over 50 percent of all ex-addict graduates either go out and establish new communities or are reemployed within similar programs relating to addiction.

The most successful ex-addicts within the communities are those who are articulate and capable of assuming leadership. In the beginning there was very little emphasis on education within the community because it interfered with the therapeutic process. In recent years, with some entire families becoming permanent residents, some therapeutic communities have become strongly education-oriented. Synanon operates its own elementary and secondary schools and encourages colleges and universities to hold classes in its facilities.

The black and Hispanic communities have criticized many of the therapeutic community programs because of the personality and identity destruction that has accompanied encounter therapy. These people feel they have been stripped of their identity by white society for too long, and that the reconstruction of black and Hispanic identity must be a valid concern for any such therapeutic program. Thus an increasing number of rehabilitation programs, especially in larger cities, now have Hispanic and black counselors.

RETURN TO SOCIETY

Both the therapeutic communities and the maintenance programs tend to reduce compulsive drug abuse and addiction to the level of an individual problem. This may well be the key to drug-abuse treatment. After an individual has been treated for drug abuse and has returned to society, many personal (social, legal, economic, and medical) problems, may be faced. The offender and the victim may be the same person, and the social pressures that encouraged abuse in the first place are still present in this environment upon return.

If there are facilities available to provide for a gradual reentry into society, or if there has been no criminal record (civil commitment rather than criminal conviction) established because of drugs, the individual has a much better chance of adjustment. Short visits home should be made at first, then a halfway house, work camp, parish house, or a day–night hospital may help beyond the third phase of the therapeutic community. Any of these settings is potentially useful in providing the abuser with social, therapeutic, educational, and vocational services; these provide controlled contacts with the community.

But all too often, the treated drug abuser leaves the hospital (or more often, the jail) and is literally dumped back on the street. Consequently, it is a very short time before the person is again abusing drugs and is back to or worse off than when treatment was first initiated.

If the individual has been convicted of a felony, there may be further problems. Such a conviction can mean the loss of voting rights, the loss of the right to hold public office, and the loss of the right to obtain certain state licenses such as those to sell liquor, practice law, or dispense drugs.

Such a felony conviction can also hamper the person economically. Very often an individual convicted of a felony may be required to register with the local police department of a city or area, or may be required to register with the emigration agency whenever leaving or entering this country and other countries. The person will likely be unable to be bonded and to hold positions that require a bond. Because of this and the inability to acquire types of security clearances through the government, the person is also excluded from certain areas of employment such as defense establishments or defense contractors. Many employers refuse to hire a person with a record of felony conviction.

TOWARD THE FUTURE

Drug-treatment programs can only succeed when they deal with the basic fact that the compulsive user or addict has no ability to combat the ordinary stresses of life. The abuser has relied for months or perhaps years on the external solace provided by drugs. Any attempt at a cure for the addiction must help redirect the abusers' attitudes toward their own weaknesses.

The cure of the confirmed drug user is simply not a usual prospect. A great deal more research will be necessary in order for society to discover ways to maintain a consistent level of cure and rehabilitation. The damage done by drug abuse is so powerful and so widespread that the only practical long-range cure is prevention.

Obviously, the trend is away from purely punitive measures and toward rehabilitation and treatment. There is no doubt that rehabilitation and planned prevention hold much greater promise for both the individual and society than punitive confinement does.

A person who has been rescued from a possible life of drug abuse has saved society a good deal more than the cost of treatment. The more this nation's policy makers become convinced that the cost of ambitious, well-planned, preventive programs is money well spent, the more progress we can expect to witness.

SUMMARY

I. Philosophies of drug-abuse control
   A. One sees a drug offender as a person whose criminality manifests itself in drug offenses.
   B. Another sees a drug user as a sick person who must be protected from his or her self and be prevented from spreading the disease to others.
   C. Most drug-control laws are directed toward the criminality of the person, rather than toward his or her illness.

II. Legal controls of drug abuse
   A. Legal controls include quarantines, isolation, and penalties.
   B. Most laws are concerned with the illegal possession, manufacture, and sale.
   C. Federal laws.
      1. Pure Food and Drugs Act (1906)—required food and drug manufacturers to list amounts of alcohol and habit-forming drugs on labels.
      2. Harrison Narcotics Act (1914)—restricted importation, manufacture, sale, and dispensing of narcotics.
      3. Narcotic Drugs Import and Export Act (1922)—authorized importation of specific quantities of crude opium and coca leaves for medical and scientific needs.
      4. Marijuana Tax Act (1937)—required the registration and payment of a tax for importing, manufacturing, producing, compounding, selling, dealing in, dispensing, prescribing, administering, or giving away marijuana.
      5. Food, Drug, and Cosmetic Act (1938)—divided drugs into prescription drugs and over-the-counter drugs.
      6. Opium Poppy Control Act (1942)—outlined requirements for obtaining a license from the federal government for the cultivation of opium in the United States if and when needed.
      7. Boggs Act (1951)—spelled out penalties for violating federal narcotics laws.
      8. Durham-Humphrey Act (1951)—made dispensing of barbiturates and amphetamines without a prescription illegal.
      9. Narcotic Control Act (1956)—increased sentences at all levels for drug offenses.
      10. Drug Abuse Control Amendments (1965)—established legal definition of a "dangerous drug."
      11. Narcotic Addict Rehabilitation Act (1966)—changed status of two federal narcotics hospitals from treatment centers to research centers.
      12. Comprehensive Drug Abuse Prevention and Control Act (1970)—provided new schedule of federal drug penalties to replace and update older drug laws.
         a. This law outlined five classes of drugs and penalties for violation of each.
         b. Class 1 drugs have the highest potential for being abused, and Class 5 drugs have the lowest potential for abuse (Table 5.1).
   D. State and local laws.
      1. Most police departments in large cities have separate narcotics units.

SUMMARY

2. Most states have now adopted versions of a model uniform state narcotic law patterned after the Harrison Narcotics Act.
3. Yet state laws dealing with dangerous drugs are anything but uniform.
4. Many states are currently in the process of remodeling their drug laws.
5. California requires a triplicate prescription for the dispensing of any narcotics—one copy stays with the physician, one copy with the pharmacist and one copy goes to the state Bureau of Narcotics (Figure 5.1).

III. Drug control and law enforcement
   A. There is still a lack of consensus today on drug control in the United States.
   B. The debate over legalization of marijuana, as a case in point, is evidence of the ambivalent feelings as to how it should be controlled.
   C. Nine ways in which wrong attitudes toward the problem of drug abuse develop (Helen H. Nowlis, University of Rochester):
      1. Ignorance of the actions of chemical substances in living persons.
      2. Semantics that become entangled in myths, emotions, assumptions, and beliefs.
      3. Communication that fails to develop between scientists and the general public.
      4. Lack of understanding of the scientific method and concepts.
      5. Living in a world of constant and dramatic change has contributed to a disorientation of the traditional beliefs.
      6. The philosophy of social control in place of unenforceable laws has been difficult for many people to understand and accept.
      7. Education, although an essential key to understanding, has been the target of shortsighted attacks upon it.
      8. American society is a pill society—a fact that has made the solutions of drug abuse difficult.
      9. Increasing retreat in the face of complex problems has characterized too much of society.

IV. Drug-abuse treatment
   A. Needed control may be legal, social, or self-control.
   B. The ultimate success of any control will depend upon the motivation of the abuser toward behavior modification that leads toward an independent, productive existence.
   C. Treatment laws
      1. Narcotic Addict Rehabilitation Act of 1966 gave federal

support through the National Institute of Mental Health to develop drug-treatment programs within the individual states.
2. Many states subsequently passed legislation supplementing these laws. Figure 5.2 outlines the procedural laws enacted in California.
3. Civil Commitment for Drug Abuse is a mechanism utilized to ensure control over drug abusers during rehabilitation, without the drug abuser establishing a criminal record when seeking help.
4. Criminal Commitment for Drug Abuse is commitment by a judge to a treatment program for an indeterminate period of time.

D. Drug-abuse therapy
1. A behavioral continuum exists along which would be found users from the regular user to the compulsive user, or addict.
2. Any therapy program must be directed toward the troubled person, rather than toward the drug.

E. Modes of therapy
1. Tests for drugs
   a. The Nalline test (1955) was one of first drug tests for evidence of use of drugs.
   b. The free radical assay technique (FRAT) measures free and bound morphine in the urine.
   c. The thin layer chromatography (TLC) test detects the presence of a number of drugs.
   d. The gas liquid chromatography (GLC) test is another such test.
2. Initial emergency therapy
   a. Often first treatment is rendered because of the acute condition of the patient.
   b. Often called the talk-down phase, its success may depend upon the degree of rapport established between physician and patient.
3. Withdrawal therapy
   a. Withdrawing the patient from the drug may be complicated and abrupt withdrawal can be fatal.
   b. If withdrawal is successful, the user may have sufficient motivation to give up the use of illicit drugs.
4. Narcotic-antagonist therapy
   a. Narcotic antagonists are drugs so chemically and structurally similar to true narcotics, the nervous system can not distinguish between the two.
   b. The narcotic does not provide the abuser with the narcotic effect, but rather will, when given to an addict, bring on withdrawal symptoms.

SUMMARY

   c. To be fully effective, the use of antagonists must be accompanied by a comprehensive program of psychological and social rehabilitation.
  5. Drug maintenance therapy
   a. This form of therapy involves the temporary continuation of drug substitutes that resemble illicit drugs but which block the euphoric effects of opiates.
   b. One such drug substitute is *methadone,* whose use is still considered controversial.
  6. Therapeutic communities
   a. Complex social communities, houses, or social systems are designed to provide care through encounter-group therapy, highly structured community organization, and a reward-punishment system based on simple behavioral psychology.
   b. Such communities are designed to develop and reinforce standard middle-class norms of behavior and attitudes. Reentry of the person into the community is the expectation.
   c. To the extent that therapeutic committees fail to lead the patient back into the community, they represent some degree of failure.

V. Return to society
 A. Both the therapeutic communities and the maintenance programs tend to reduce compulsive drug abuse and addiction to the level of an individual problem.
 B. Gradual reentry of the individual into society provides the best chance of adjustment. This can be facilitated through a halfway house, work camp, parish house, or a day–night hospital.
 C. A felony conviction of the patient adds further problems and can definitely hamper reentry into society through the loss of many normal privileges.

VI. Toward the future
 A. The success of drug-treatment programs depends upon their recognizing that the compulsive user or addict has no ability to combat the ordinary stresses of life.
 B. Since present programs of rehabilitation have been so unsuccessful, the only practical long-range cure is prevention.
 C. Punitive measures offer no significant promise for the future.
 D. A person rescued from a life of drug abuse has saved society much more than the cost of treatment.

6

ALCOHOLIC BEVERAGES: USE AND ABUSE

Alcohol use has many different facets. Most North Americans find it a pleasant and generally enjoyable part of dinner parties, social gatherings, celebrations, and so on. Unlike the use of hallucinogens and opiates, the use of alcohol in North American society is not necessarily questioned or condemned; nor is it illegal. It becomes a problem for individuals and society only under specific conditions: driving while under the influence of alcohol, public intoxication, the tragic personal consequences of excessive drinking. The effects of alcohol are part of a complex web: some are definitely a result of problems of body chemistry, and some relate more to social control and responsibility. Often severe psychological problems are involved. In this respect, alcohol has many similarities to the other drugs discussed in this book. It is difficult to predict all its effects or to understand how it affects different individuals.

Ironically, because of the generally freer attitudes toward alcohol use, it is possible to see and understand more of the harmful results of its overuse. With the possible exception of heroin, more is known about the damage done by alcohol than about any other drug discussed in this book.

At the same time that alcohol abuse is such a widespread problem, millions of people derive great enjoyment from imbibing alcoholic beverages without ever seriously threatening either their personal or society's well-being. The reasons for the differences between socially acceptable and socially unacceptable mood-modifying drugs are not all clear, but here and in Chapter 7 we will consider those facts about alcohol that are currently known to science and medicine.

A SOCIALLY ACCEPTABLE MOOD-MODIFYING DRUG

The main distinction between alcohol and many of the substances discussed in the previous chapters is that the moderate use of alcohol is legally and socially acceptable within the majority of North American subcultures. About 70 percent of all adults in North America make at least some use of alcoholic beverages. Yet alcohol has a potential for abuse as great as that of any of the previously discussed drugs.

In Chapter 3 alcohol was identified as one of the mood-modifying drugs: a *sedative*. It is the most widely used and abused mood-modifying sedative in North America. For that reason Chapters 6 and 7 will be devoted to a discussion of alcoholic beverages and alcoholism (dependent alcohol abuse or alcohol addiction).

ALCOHOLIC BEVERAGES

In order to understand the effects of alcoholic beverages, it is necessary to know something about their nature. Even among regular drinkers, there are many misconceptions regarding alcohol and alcoholic beverages.

Types of Alcohol

Although there are many kinds of alcohols, the principal one of importance in alcoholic beverages is *ethyl alcohol,* known chemically as *ethanol* and commonly as *grain alcohol*. Another common alcohol is *methyl* alcohol, usually referred to as *wood alcohol*. Methyl alcohol, which is used in many products such as antifreeze and fuels, is a deadly poison; even small amounts can cause blindness or death. Bootleg liquor (sold illegally, without payment of taxes) is occasionally found to contain methyl alcohol, and it has been the cause of many deaths. A third common alcohol is *isopropyl alcohol,* the principal ingredient of most rubbing alcohols. Although it is not as deadly as wood alcohol, rubbing alcohol is definitely too poisonous to be consumed as a beverage.

Throughout the remainder of this chapter, the unqualified term "alcohol" will indicate ethyl alcohol. Such alcohol in any alcoholic beverage is produced by the fermentation of sugar by yeasts. Each type of alcoholic beverage is produced from a specific source of sugar (Table 6.1). Beer and ale, for example, are made by fermenting malted (sprouted) barley. Wine is fermented grape juice. The "hard" liquors are made from the distilled product of the fermentation of grains and other sugar sources. Because distillation greatly concentrates the alcohol content of a beverage, the distilled liquors are usually diluted with water, soft drinks, fruit juices, or other mixes rather than being con-

TABLE 6.1
Some Common Alcoholic Beverages

Beverage	Product Fermented	Distilled?	Percent of Alcohol by Volume	Alcohol Content by Proof
Beer	Malted barley	No	4–6	8–12
Ale	Malted barley	No	6–8	12–18
Wines				
Dry (dinner)	Grape juice	No	12–14	—
Sweet (dessert)	Grape juice[a]	No	19–21	—
Whiskey	Malted grains	Yes	40–50	80–100
Brandy	Grape juice	Yes	40–50	
Rum	Molasses	Yes		80–100
Vodka	Potatoes and other sources[b]	Yes	40–50	80–100
Gin	Various sources[c]	Yes	40–50	80–100

[a] The sweet wines have sugar added after fermentation and are fortified by the addition of brandy to kill the yeast and prevent fermentation of the sugar.
[b] Vodka is essentially just alcohol and water, without other flavoring agents.
[c] Gin is flavored with extracts from juniper berries and other sources.

sumed "straight." The alcohol content of distilled beverages is expressed as "proof," which in the United States is the percent of alcohol multiplied by two. For example, 100 proof means 50 percent alcohol.

Other Ingredients of Alcoholic Beverages

In addition to alcohol and water, drinks contain varying amounts of flavoring and coloring agents collectively called *congeners*. These chemicals contribute to "hangovers" and other toxic effects of alcoholic beverages. Alcoholic beverages have almost no food value. As shown in Table 6.2, there are no vitamins, minerals, fats, proteins, or usable carbohydrates in alcoholic beverages. The one exception is beer, and the amounts present are nutritionally insignificant.

Calories, however, are abundant in all alcoholic beverages. Most of the caloric value of alcoholic beverages is derived from the alcohol itself. These are empty calories; that is, they provide nothing toward good nutrition but displace potentially nutritious foods from the diet. Alcoholics often suffer from malnutrition, and many of their physical ailments are believed to be aggravated by this condition.

Thus, alcohol can be described as a mood-altering drug with a significant caloric value. As a food, alcohol can be consumed with meals and will go a long way toward providing one's daily calorie requirement. But as a drug, alcohol can have serious effects on the normal functioning of the mind and body.

TABLE 6.2
Nutritional Values of Alcoholic Beverages

Beverages	Total Calories	"Empty" Calories from Alcohol	Nutritional Usable Calories	Protein (grams)	Fat (grams)	Carbo-hydrates (grams)	Thiamine (B_1) (mg)	Nicotinic Acid (Niacin) (mg)	Riboflavin (B_2) (mg)
Beer (12 ounces)	171	114	57	2 (8 Kcal)	0	12 (49 Kcal)	0.1	0.75	10
Whiskey (2 ounce "shot")	140	140	0	0	0	0	0	0	0
Wine (8 ounce glass)	275	240	35	0	0	8.5 (35 Kcal)	0	0	0

Figure 6.1. Equivalent drinks. The three drinks shown are roughly equivalent in their alcohol content. About the same amount of alcohol is contained in 12 ounces of beer, 4 ounces of dry wine, or 1–1 1/4 ounces of distilled liquor.

EFFECTS OF ALCOHOL ON THE HUMAN BODY

Alcohol's effects can be viewed from either the short range or the long range. A person who is neither alcoholic nor a heavy drinker will go through a definite pattern of physiological and psychological experiences when he or she has something to drink. If such a person has too much to drink, he or she might also go through a period of discomfort, a "hangover," the next day and then will return to normal, feeling only slightly worse for wear. In all probability, there will be no permanent effects.

All alcoholic beverages have basically the same effects on the body. The only important difference among them is the amount of alcohol they contain. Any two drinks containing the same amount of alcohol will produce the same effect. Table 6.3 and Figure 6.1 compare the amounts of several different drinks that contain similar qualities of alcohol.

Blood-Alcohol Concentration

The best quantitative measure of what is happening physiologically as a normal, healthy person drinks is the blood-alcohol concentration. After a drink, alcohol shows up in the bloodstream very quickly. At first,

EFFECTS OF ALCOHOL ON THE HUMAN BODY

TABLE 6.3
Amounts of Different Alcoholic Beverages Yielding Similar Quantities of Alcohol

Beverage	Alcohol Content (percent)	Approximate Amount Yielding 0.5 ounces of Alcohol (ounces)
Beer	4	12
Dinner wine	12	4
Dessert wine	21	2.5
80 proof liquor	40	1.25
100 proof liquor	50	1

in small amounts, alcohol is absorbed into the blood through the lining of the stomach; but this process slows and then stops just as quickly. The presence of food in the stomach slows absorption. This is the reason many people prefer some food at cocktail parties; it helps modify the effects of the alcohol. Unlike the stomach, the intestinal lining absorbs alcohol rapidly, regardless of the amount of alcohol or the presence of food. Consequently, moderate and high blood-alcohol concentrations are controlled by the emptying time of the stomach. Anything that slows the emptying of the stomach also reduces the blood-alcohol concentration by reducing the rate at which the alcohol is absorbed by the circulatory system. Carbon dioxide (CO_2) has the opposite effect; it speeds up the passage of alcohol into the intestines. Thus, a carbonated mixer with whiskey is a more rapidly potent drink than a whiskey-and-water highball. Dissolved CO_2 is what gives champagne its extra kick.

Table 6.4 shows the relationship between body size, number of drinks, and resultant blood-alcohol concentration. Body size is a factor because the larger the bloodstream into which the alcohol passes, the more diluted it will be. In addition to total body weight, the relative amount of blood and other fluids in the body influences a person's degree of intoxication. Women tend to have less blood volume and total body fluid than men of similar body weights. This allows less dilution of the alcohol in women. Thus, the same amount of alcohol results in a higher blood-alcohol level and a greater degree of intoxication in women than in men of the same weight.

Of even more importance is the difference in degree of intoxication possible among individuals at the same blood-alcohol level. Part of this is due to inherent biochemical differences among individuals, part to psychological differences, and part to the degree of alcohol tolerance each person has developed. Frequent heavy drinkers are definitely less affected by given amounts of alcohol than are more moderate drinkers.

Effects of Alcohol on the Central Nervous System

The most important effect of alcohol is its depressant, sedative or, in high concentrations, anesthetic action on the brain. Despite years of

TABLE 6.4
Blood-Alcohol Levels (Percent Alcohol in Blood)

Body Weight	\multicolumn{12}{c}{Drinks[a]}												
	1	2	3	4	5	6	7	8	9	10	11	12
100 lbs.	.038	.075	.113	.150	.188	.225	.263	.300	.338	.375	.413	.450
120 lbs.	.031	.063	.094	.125	.156	.188	.219	.250	.281	.313	.344	.375
140 lbs.	.027	.054	.080	.107	.134	.161	.188	.214	.241	.268	.295	.321
160 lbs.	.023	.047	.070	.094	.117	.141	.164	.188	.211	.234	.258	.281
180 lbs.	.021	.042	.063	.083	.104	.125	.146	.167	.188	.208	.229	.250
200 lbs.	.019	.038	.056	.075	.094	.113	.131	.150	.169	.188	.206	.225
220 lbs.	.017	.034	.051	.068	.085	.102	.119	.136	.153	.170	.188	.205
240 lbs.	.016	.031	.047	.063	.078	.094	.109	.125	.141	.156	.172	.188

Under .05 Driving is not seriously impaired.

.05 to 0.10 Driving becomes increasingly dangerous. 0.8 legally drunk in Utah.

.10 to .15 Driving is dangerous. Legally drunk in many states.

Over .15 Driving is *very* dangerous. Legally drunk in any state.

[a] One drink equals 1 ounce of 100 proof liquor or 12 ounces of beer.
Source: Reprinted through the courtesy of the New Jersey Department of Law and Public Safety, Division of Motor Vehicles, Trenton, New Jersey.

EFFECTS OF ALCOHOL ON THE HUMAN BODY

research, the exact way in which alcohol intoxicates is still not definitely known. Its depression of the central nervous system is progressive and continuous. As shown in Figure 6.2, the higher centers (cortex) are depressed first. As the blood-alcohol concentration increases, the depression continues down the central nervous system.

Alcohol is a mood-modifying drug and can temporarily produce a state of mild euphoria and an apparent stimulation. This is usually the basis of its attraction for the social drinker (occasional user). The deeper depression, which lets a person escape from cares, pressures, tensions, and anxieties, that takes place at higher concentrations of blood-alcohol levels (Table 6.5) is what the heavy (abusive) drinker is seeking. The stimulant effect of alcohol is an illusory one, however.

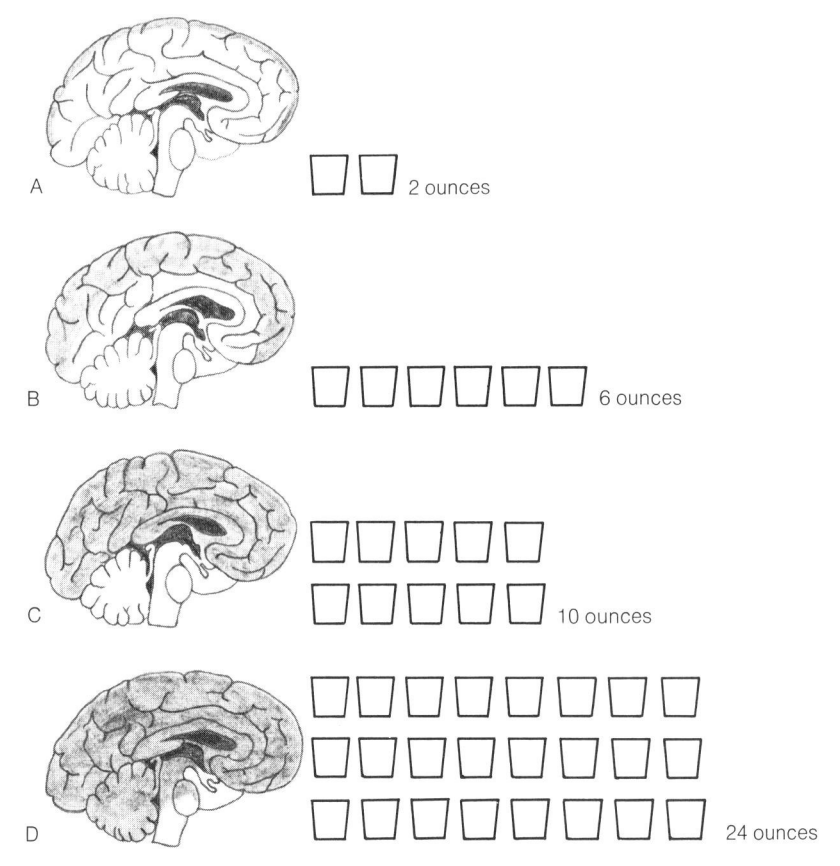

Figure 6.2. Alcohol and its effects on the brain: (A) Higher brain centers affected. (B) Deeper, motor areas affected. (C) Emotional centers of the midbrain affected. (D) Entire perception area affected. [Modified from Betty S. Bergersen and Andres Goth, Pharmacology in Nursing, *12th ed. (St. Louis: The C. V. Mosby Co., 1973)], p. 281, fig. 13-4.*

TABLE 6.5
Some Physiological Effects of Alcoholic Beverages

Amount of Alcoholic Beverage Consumed	Concentration of Alcohol Found in the Blood (in %)	Physiological Effects	Time Required for Alcohol To be Oxidized (hours)
1 highball (1½ oz. whiskey) 1 cocktail (1½ oz. whiskey) 3½ oz. fortified wine 5½ oz. wine or pop wine 2 bottles beer (24 oz.)	0.03	Slight changes in feelings.	2
2 highballs 2 cocktails 7 oz. fortified wine 11 oz. wine or pop wine 4 bottles beer	0.06 (0.08—legally drunk in Utah)	Increasing effects with variation of individuals and within the same individual at different times.	Feelings of warmth, emotional relaxation, slight decrease in fine skills. Less concern with minor irritations and social restraints.	4
3 highballs 3 cocktails 10½ oz. fortified wine 15½ oz. (1 pt.) wine or pop wine 6 bottles beer	.09 (0.10—legally drunk in many states)	Emotional buoyancy, exaggerated emotions and behavior. Person talkative, noisy or morose.	6
4 highballs 4 cocktails 14 oz. fortified wine 22 oz. wine or pop wine 8 bottles beer (3 qts.)	0.12	Impairment of fine coordination, clumsiness. Slight to moderate unsteadiness in standing or walking. Loss of peripheral vision.	8
5 highballs 5 cocktails 17½ oz. fortified wine 27½ oz. wine or pop wine ½ pint whiskey	0.15 (legally drunk in all states)	Intoxication, unmistakable abnormality of body and mental control.	10

Note: Based upon individual of "average" size (150 pounds). See Table 6.4, Blood-Alcohol Levels (Percent Alcohol in Blood), to see effects on individuals of varying weights.

EFFECTS OF ALCOHOL ON THE HUMAN BODY

Actually, alcohol slows down the functions of the brain and central nervous system. As Table 6.5 indicates, the first part of the brain to go is the center that controls judgment and inhibitions. Thus, paradoxically, alcohol stimulates drinkers for a brief period by depressing the restraining factors in the cerebral cortex. Many drinkers become talkative and happy and assume that they are being witty and charming. Frequently, at this stage, the stimulation will cause someone to say and do things that they would, if sober, prefer left unsaid and undone.

Alcohol interferes with both the storage and the retrieval of information. When people are under the influence of alcohol, their ability to learn and to recall past events and information is decreased. Problem-solving ability is also greatly diminished. Even simple puzzles and arithmetic problems may be difficult or impossible for the intoxicated person to solve.

Recent research (Jones and Jones, 1976) has shown that the period of alcohol effects is composed of a sequence of four distinct, obvious, and clearly definable states of consciousness. Alcohol State of Consciousness-1 (ASC-1) is identified with the period of alcohol absorption and increasing blood-alcohol levels. This is the time of stimulantlike effects such as talkativeness and laughter, deteriorating muscle coordination, and declining performance in mental and physical tasks.

Alcohol State of Consciousness-2 begins after the peak blood-alcohol level has been obtained and the blood alcohol level is declining. The person becomes quiet and tired, but the alcohol effects are still apparent to the individual and to observers. Mental and physical performance is still impaired but is beginning to improve.

Alcohol State of Consciousness-3 is less well defined than ASC-1 or ASC-2 but begins to appear about halfway between the peak blood-alcohol levels and zero blood-alcohol level. ASC-3 is the time when the individual feels confident that he or she is perfectly sober and no longer under the influence of alcohol. Yet, a detectable blood-alcohol level is still present. The person feels sober, but does not act sober. Reaction time and muscle coordination are still impaired. This is probably the most dangerous alcohol state. While people in ASC-1 and ASC-2 realize that they are intoxicated, the person in ASC-3 does not realize it and probably will not admit that his or her driving or other performance is impaired. People in ASC-3 know what they want to do but just cannot do it properly. Their thoughts are dissociated with their actions, and even the evidence of their poor performance may not convince them that they are not sober.

Alcohol State of Consciousness-4 follows the return to a zero blood-alcohol level. Most people would say that a person is no longer under the influence of alcohol from this point on. However, experimental evidence does not support this assumption. For example, certain abnormal eye movements called *positional alcohol nystagmus* can be recorded for up to 24 to 32 hours after drinking. Such eye movements may be sufficient to impair one's ability to fly an airplane or perform similar critical tasks. Thus, some individuals may unknowingly be vulnerable to effects from alcohol for more than a day after they first consume the alcohol.

Many people worry about possible permanent effects on the brain as a result of drinking. Light or moderate drinking apparently has little or no permanent effect on the brain. However, if a person drinks long enough and in sufficient quantities, permanent brain damage can occur. This damage is discussed in Chapter 7.

Effects of Alcohol on Other Organs

In addition to its effect on the brain, alcohol directly and indirectly affects some of the other body organs.

Alcohol and the skin. One easily noticed effect of alcohol is the dilation of the small blood vessels of the skin. As a result, the face and neck appear red, and the person feels warm. Rather than an effective defense against cold weather, however, the use of alcoholic beverages as a means of keeping warm is actually dangerous. Body heat may be lost to the point that internal body temperature drops to a dangerously low level.

In alcoholic persons small blood vessels on the nose, cheeks, neck, and chest may remain dilated, even in the absence of alcohol. To a physician, these dilated vessels may provide the first evidence of a patient's alcoholism.

Alcohol and the senses. Sight is the first sense affected by alcohol. Although small amounts of alcohol increase sensitivity to light, there is decreased ability to distinguish between two different intensities of light, and focusing becomes more difficult. Increasing the amount of alcohol causes a great loss of vision. Hearing is affected less than sight but is still significantly impaired at higher blood-alcohol levels.

Alcohol and the muscles. Because all the voluntary muscles are under the control of the brain and nervous system, muscle control is impaired at all blood-alcohol levels. This results in a loss of coordination and a lengthened reaction time, and, of course, these changes are especially detrimental to automobile drivers.

Alcohol and the digestive system. There are several possible effects of alcohol upon digestion and the digestive system. Moderate drinkers sometimes notice an improvement in their digestion when they drink before or with a meal. One explanation for this is that alcohol can indirectly improve digestion by relieving the nervous tension which may interfere with digestion. Alcohol may also aid digestion by stimulating the production of stomach acid. But large amounts of alcohol irritate the stomach lining and, along with the increased amount of acid, may cause *gastritis* and other stomach conditions. High concentrations of alcohol in the stomach may cause enough irritation to trigger the reflex action of vomiting to remove the irritant.

Alcohol and the liver. No tissue in the body escapes some damage from heavy alcohol use, but liver damage is the major cause of disabling illness and premature death in alcoholics. Alcoholic liver disease begins by the accumulation of excess fat in the liver, so-called *fatty liver*. This is a very common complication of heavy drinking, but it is usually reversible if the drinking ceases. If the drinking continues, the liver may become further damaged, resulting in *cirrhosis*. Cirrhosis is characterized by scarring of the liver, increased fibrous tissue, distortion of the normal anatomy of the liver, and alteration of its normal functioning. This damage is irreversible and frequently results in death. The damage obstructs blood flow through the liver, causing hypertension (high blood pressure) in the *portal vein,* accumulation of fluid in the abdomen (*ascites*), and gastrointestinal bleeding. Such bleeding and other complications such as *coma,* which is due to reduced liver function, are what lead to its fatal outcome. In large cities, cirrhosis represents the third or fourth leading cause of death in males.

Though not all people affected by cirrhosis are alcoholic, the majority are heavy drinkers. For many years the relationship between alcoholism and cirrhosis was a subject of debate. One of the most hotly debated issues dealt with whether this liver disease in alcoholics was due to alcohol itself or to the malnutrition so commonly associated with alcoholism. It is now clear that adequate diets do not prevent alcohol from producing either fatty liver or cirrhosis. The damage is caused by alcohol, not by malnutrition. Adequate nutrition is essential for the normal functioning of all organs, including the liver. Alcoholic persons have a tendency to eat poorly and often do not absorb essential nutrients, even when they are present in the diet. Combining such malnutrition with alcoholism is obviously adding insult to injury for the liver and other organs.

Alcohol and the heart. The myth still persists that alcohol has a beneficial effect on the heart. However, studies (Regan, 1793/74) have shown that, even in moderate amounts, alcohol can adversely affect heart function. In intoxicating amounts, it can depress heart function severely and damage the heart muscle fibers. It appears that the cumulative effect of frequent heavy drinking can produce structural and functional abnormalities of the heart. Effects may eventually include heart failure, abnormal heart rhythms, or the development of blood clots. Furthermore, these effects have been shown to occur in individuals in whom malnutrition, though common in alcoholics, is not present. Also, evidence indicates that if drinking ceases, heart function returns to normal in most individuals.

Sexual effects of alcohol. Many people report that the pleasure they derive from sexual intercourse is increased by alcohol. Similarly, many people note that they are more eager to engage in sexual activity when they are drinking. Thus, it is surprising to learn that alcohol is not an *aphrodisiac* (sexual stimulant). Alcohol is, however, a great "disinhibitor." Since many people are usually sexually inhibited, the loss of inhibitions is how alcohol has gained a reputation as a sexual stimulant. However, large amounts of alcohol can have a disastrous effect on sexual performance. Excessive drinking, particularly on a chronic basis, is a common contributor to impotence (*erectile dysfunction*) among males.

Sobering Up

Small amounts of alcohol are lost from the body with the breath, sweat, and urine; but 85–90 percent of the alcohol taken into the body is disposed of by oxidation into carbon dioxide and water. This oxidation takes place through several steps, the first of which occurs almost exclusively in the liver. Although the remaining steps take place rapidly in various body tissues, it is the liver that governs the speed of the total process, since the second step cannot take place until the liver completes the first step.

Once the liver has converted the alcohol to carbon dioxide and water, it can then be exhaled and excreted. The ability of the liver to handle this process determines the speed at which a person will "sober up." Generally, the oxidative process can handle one drink per hour; during a four-hour party, drinking at the rate of one drink per hour will usually not cause excessive intoxication.

Every individual has a rate at which he or she can oxidize alcohol. This rate varies among different people, but it cannot really be affected by such factors as drinking black coffee or walking in cold night air. Only time can "sober up" an individual.

Evidence indicates that for a long period of time, during the progression of alcoholism, prolonged drinking of more alcohol than the liver can comfortably metabolize leads to the development of a supplemental system for alcohol metabolism in the liver. Under these circumstances, the ability of the liver to metabolize alcohol may double. But this tolerance is reversed later as the liver becomes damaged, greatly reducing the liver's ability to metabolize alcohol.

The Hangover

A common experience following excessive drinking is the "hangover." A hangover consists of the headache, nausea, tiredness, shaking, dizziness, thirst, and dehydration common after drinking too much alcohol. Hangovers are due to several factors. First, *acetaldehyde,* a product of the breakdown of alcohol by the liver, causes toxic reactions in the brain. Acetaldehyde is responsible for much of the headache associated with hangovers. Second, the *congeners* (chemicals other than ethyl alcohol) present in alcoholic beverages are important causes of hangovers. They contribute to headaches, nausea, shaking, and dizziness. Each type of beverage has its own amount and type of congeners, producing its characteristic flavor. In general, darker colored beverages are higher in congeners than are lighter colored drinks. Third, the diuretic effect of alcohol itself increases the flow of urine, dehydrates the body, and produces some of the ill feeling of the "morning after." In addition, the cigarette smoke often present in bars and at parties may affect nonsmokers, causing further headache and other symptoms. Smokers often smoke more heavily than usual when drinking, building up very high levels of carbon monoxide and other toxic products in their blood.

People have used many techniques in dealing with hangovers. Most treatments are of only limited value since it is virtually impossible to increase the speed at which the body metabolizes the various toxic products causing hangovers. Drinking plenty of water to restore fluid balance is perhaps the single most beneficial thing a person can do to combat hangover.

The predictability of many of the unpleasant side effects of excessive drinking is usually instrumental in helping people control their use of alcohol. After a few years of alcohol use, most people do learn how much alcohol is enough for mild, pleasant stimulation; how certain drinks affect their ability to function sensibly and responsibly; and what combination of beverages might be harmful or unpleasant to them.

This discussion has presented a portion of what science can contribute to the subject; the task of developing sensible attitudes toward alcohol use rests on the individual. Undoubtedly, the people who really enjoy alcohol the most are those who use it properly.

PROPER USE OF ALCOHOLIC BEVERAGES

There are some people who argue that any consumption of alcoholic beverages is improper. But the majority of Americans find no medical, moral, legal, or religious reason for not making moderate use of alcoholic beverages. We shall therefore offer some suggestions that may help a person avoid problems in drinking (Table 6.6).

Even those people who fully approve of drinking and who themselves

TABLE 6.6
Avoiding Problems with Alcohol

---

1. Never drink because you feel like you "need" a drink. Find some alternative anxiety-reducing mechanism such as exercise, an enjoyable hobby, the companionship of someone you enjoy, or a movie or other diversion. Also, work on a real solution to the problem that is causing your anxiety.
2. In social situations, drink *slowly*.
3. Always consume food along with alcoholic beverages to slow the absorption of alcohol.
4. After (or instead of) a couple of drinks, switch to some nonalcoholic beverage such as fruit juices, soft drinks, or coffee.
5. If you notice any of the symptoms of early alcoholism (see page 190) developing in yourself, *stop drinking*. It is foolish and self-destructive to persist in drinking in light of these symptoms.

---

drink regularly usually disapprove of certain types of drinking behavior, for example, drunken driving or such antisocial behavior as physical or verbal violence. Almost all those who approve of drinking do feel that there are times and places where drinking is not appropriate. Any time that a person needs his or her fullest mental facilities, such as when driving or flying or operating machinery, is obviously a poor time to drink. For many employers, drinking or being drunk on the job is grounds for immediate dismissal. Other employers offer assistance to their employees in overcoming drinking problems.

It is very poor policy to drink for courage, for example, in preparation for a job interview or sales conference. This is using alcohol as a crutch and is a step in the direction of alcoholism.

There is, of course, no set answer for the question of how much to drink. While drinking is acceptable in American society, getting drunk is definitely frowned upon. The person who drinks is expected to drink in moderation, without serious impairment of physical or mental functions. The moderate drinker learns to drink slowly and to pace drinking so that a high blood-alcohol level does not result.

The host or hostess of a party at which drinks are served should feel a certain responsibility for the amount of alcohol the guests consume. How would you feel if someone was involved in a fatal accident on the way home from your party? There should be nonalcoholic drinks available for those who prefer them, and the person who prefers not to drink should not be pressured or ridiculed. There should be plenty of food available at all times during the party. No pressure should be put on any guest to drink more than he or she wants. A person who wants to stop at one drink should be allowed to. During the last hour or so of a party, the bar should be closed and coffee should be served. This serves several purposes. Coffee does not counteract alcohol, but the caffeine may help overcome the drowsiness that can be as much a cause of accidents as intoxication. The time spent drinking coffee serves as a "sobering-up" period as well. Finally, the serving of coffee is understood by most guests to be a signal that the party is about over, so the host can bring a party to a close when desired. Anyone who is obviously in no condition to drive home should be strongly encouraged to stay overnight, take a taxi, share a ride, or do something other than drive.

YOUNG PEOPLE AND ALCOHOL

Across North America more young people are drinking than ever before. Also, a significant number of these young drinkers are abusing alcohol in ways that disrupt their lives and threaten the health and safety of themselves and others. Not only are more young people drinking, but the frequency of their drinking and the amount they consume have both increased.

The reasons for this increased drinking are not clear. Some authorities speculate that earlier social maturation by today's young people leads to earlier experimentation in a variety of so-called "adult" behaviors. Parental and peer influences are shown by many authorities to be contributing factors. Because parents serve as role models for adult behavior, their attitudes and practices play a major role in determining their children's approach to alcohol. On the other hand, some drinking by young people reflects rebellion against parental standards of alcohol abstinence. Having friends who drink seems to increase the probability that a young person will drink. This is because of the increased peer pressure and availability of alcoholic drinks.

Depending on the student group surveyed and the criteria used to identify problem drinking, anywhere from 5 to 30 percent of high school students can be classified as "problem drinkers." Obviously, efforts toward prevention of problem drinking among young people need to be intensified. Among the most promising approaches to prevention are those which promote responsible decision-making and increasing healthy self-concepts in young people (see Table 6.7).

PROBLEMS RESULTING FROM ALCOHOL ABUSE

Alcohol and Society

The relationship between alcohol abuse and society is complex, with each influencing the other. Many social problems, such as divorce, unemployment, and accidents, are unquestionably aggravated by excessive drinking. On the other hand, problem drinking is often a symptom of a deeper personal or social problem rather than the basic problem itself.

Much of the effect of alcohol upon society is intangible, but there are certain readily apparent costs to society from alcohol abuse. The most obvious costs are those that can be measured in dollars, for example, the cost of alcohol-related accidents, crimes, hospitalization,

ALCOHOLIC BEVERAGES: USE AND ABUSE

Decision Making About Alcohol

1. The decision to drink or not to drink should be a personal, private decision.
2. It is not essential to drink.
3. Drinking does not indicate adult status or maturity.
4. Drinking does not indicate masculinity or femininity; it does not prove sexual adequacy.
5. Anyone choosing to drink has a responsibility not to harm himself, herself, or others.
6. Those who use alcohol should avoid drunkenness.
7. Those who drink should respect the decision of other people to abstain.
8. Those who serve alcoholic beverages to guests or customers should realize they have a responsibility to these persons.
9. Adults, by the example they set, are significantly responsible for the drinking habits of youth.
10. Alcoholism is a preventable illness and a treatable one.

and unemployment and welfare payments to problem drinkers and their families. Other, less measurable costs include broken homes, premature deaths, and loss of creativity and productivity.

Alcohol and the Family

The role of alcohol in a family is influenced by the religious, ethnic, and social affiliations of that family. As long as the drinking practices of the family members are in accord with those of the social groups to which the family belongs, drinking is usually not associated with family problems.

When drinking does become associated with family problems, it is of prime importance to determine whether the drinking is the cause of the family problems or a symptom of a deeper family or personal problem. In the past, alcohol was automatically held responsible for family poverty, divorce, child neglect, juvenile delinquency, and most other family problems. Today, problem drinking is recognized as often being just one symptom of a deeper emotional problem. However, a vicious circle often develops in which personal and family problems lead to excess drinking, which leads to further family problems, which leads to the eventual destruction of the family unit.

Alcohol and Crime

The police spend a great deal of their time and effort in handling problem drinkers. Many of these people are charged only with being drunk in a public place or with being drunk and disorderly. But others have become involved in much more serious offenses, such as assault, murder, or felony traffic violations. A person need not be an alcoholic or even a regular drinker to become involved in serious trouble as a result of drinking. Many serious crimes and fatal traffic accidents have resulted from a single, isolated occasion of heavy drinking. The alcohol-related crime is likely to be an impulsive act of aggression against another person, without motive or profit.

Alcohol and Driving

Some people feel that their driving ability is improved by small amounts of alcohol, but the truth is that alcohol only makes these people think they are driving better. The drinking driver seldom realizes how much his or her driving ability has deteriorated because the same effects on the brain that cause one to be a dangerous driver also make one unaware of how poor his or her driving has become.

At low blood-alcohol levels, the main effect on driving is a reduction in the level of judgment and care used. You can still drive straight enough, but you may take chances you might otherwise not risk. Blood-alcohol concentration levels are the basis of drunk driving determinations in almost all states. Alcohol is believed to be a contributing factor in 25–50 percent of all fatal traffic accidents. Although alcohol is not officially listed as the actual cause in all of these cases, it is believed that many accidents blamed on "high speed" or "failure to negotiate a curve" are actually due to excessive drinking. Research has shown that alcohol starts to be a factor in accidents at blood levels as low as 0.03 percent (Table 6.5). With higher blood-alcohol levels, there are the additional factors of poor vision and slowed muscular reactions (reflexes).

Drunk driving is predominantly a male behavior pattern. About 98 percent of persons convicted of driving while intoxicated are males (*Special Report on Alcohol and Health,* National Institute on Alcohol Abuse and Alcoholism, 1974). A greater percentage of women than men will refuse to drive after drinking. Perhaps this relates to male ego needs and our unfortunate tendency to equate being able to drink large quantities of alcohol with "masculinity."

Furthermore, it is usually not the moderate or "social" drinker who is arrested or has an accident as a result of drunk driving. Many accidents and arrests involve people who have a history of repeated alcohol-related offenses or show numerous indications of being problem drinkers. Unfortunately, no effective way has been found to keep such drivers off the road. Many will continue to drive even after their license is revoked.

Alcohol and Life Span

Moderate drinking has not been shown to adversely affect the length of life. In fact, numerous studies (Chafetz, 1974) have shown that moderate drinkers actually live somewhat longer than abstainers. The reasons for this are not yet known.

But heavy drinkers die much earlier than either moderate drinkers or abstainers. The death rate from all of the following causes is increased for heavy drinkers: cancer of the upper digestive and respiratory tracts, pneumonia, cirrhosis of the liver, peptic ulcers, heart disease, suicide, and accidents including vehicles, falls, fires, and poisoning.

The mode of dying of heavy drinkers reflects their drinking behavior, their emotional states, their tendency to smoke heavily, and their neglect of proper nutrition and hygiene. In short, their style of living is reflected in their style of dying.

SUMMARY

I. Alcohol is a socially acceptable mood-modifying drug.
   A. The main distinction between alcohol and other drugs is legal and social acceptability.
   B. It is the most abused drug in North America.

II. Alcoholic beverages
   A. Types of alcohol:
      1. Ethyl or grain alcohol—beverage alcohol
      2. Methyl or wood alcohol—poisonous
      3. Isopropyl or rubbing alcohol—poisonous
      4. "Proof" is percent alcohol multiplied by 2 (100 proof is 50 percent).
   B. Other ingredients of alcoholic beverages
      1. Various chemicals collectively called *congeners*
      2. High caloric content
      3. Little food value

III. Effects of alcohol on the human body
   A. Blood-alcohol concentration
      1. Best available measure of what is happening physiologically as a person drinks.
      2. Depends on the amount of alcohol consumed, body weight, stomach contents at the time of drinking, and the presence of carbonation in the beverage.
      3. Women of the same weight as men have less blood volume and achieve higher blood-alcohol levels for same amount of alcohol.
      4. Even at the same blood-alcohol level, there are different degrees of intoxication due to individual biochemical differences and degrees of tolerance.
   B. Effects of alcohol on the central nervous system
      1. The most important effect is depressant or sedative action.
      2. Effects can be divided into four sequential stages called "Alcohol States of Consciousness (ASC) 1–4."
         a. ASC-1: period of increasing blood-alcohol levels; stimulantlike effects are shown.

SUMMARY

181

- b. ASC-2: blood-alcohol level is declining; person is quiet and feeling tired.
- c. ASC-3: occurs at about halfway down from peak blood-alcohol level; person feels sober, but does not act sober.
- d. ASC-4: follows return to zero blood-alcohol level; residual effects can be recorded for up to 32 hours after drinking.
   3. Light drinking causes little or no permanent brain damage; heavy drinking may have this effect.
C. Effects of alcohol on other organs
   1. Alcohol and the skin
      a. Dilation of small blood vessels that can result in dangerous loss of body heat in cold weather.
      b. Small blood vessels in the skin of alcoholic persons may be permanently dilated.
   2. Alcohol and the senses—all senses are impaired.
   3. Alcohol and the muscles—control by the nervous system (coordination) is impaired at all blood-alcohol levels.
   4. Alcohol and the digestive system
      a. Moderate drinking may stimulate digestion.
         (1) Relaxing effect
         (2) Increased production of stomach acid
      b. Heavy drinking may cause gastritis and other problems.
   5. Alcohol and the liver
      a. Liver damage is a major cause of illness and death in alcoholic persons.
      b. Early damage is called *fatty liver*.
      c. Later damage is *cirrhosis*—scarring of liver.
      d. This damage now known to be caused by alcohol, not by malnutrition.
   6. Alcohol and the heart
      a. Even moderate amounts of alcohol can adversely affect heart function.
      b. Heavy drinking can produce structural and functional changes in the heart muscle.
      c. The heart usually returns to normal if drinking is stopped.
   7. Sexual effects of alcohol
      a. It is not a true aphrodisiac.
      b. It increases the sexual pleasure of many people through release from inhibitions.
      c. Excessive drinking is a common contributor to impotence (erectile dysfunction) in males.
D. Sobering up
   1. Most alcohol is oxidized into carbon dioxide and water.
   2. The liver determines the speed of sobering up.
   3. Average rate of oxidization is about one drink per hour.

E. The hangover
   1. Hangover is due to:
      a. Acetaldehyde produced by the body in breaking down alcohol.
      b. The congeners in alcoholic beverages.
      c. Dehydration from the diuretic effect of alcohol.
      d. The cigarette smoke often present in drinking scenes.
   2. There is no really effective treatment; drinking plenty of water may help.

IV. Proper use of alcoholic beverages
   A. Guidelines
      1. Make sure the use of alcohol improves social relationships rather than impairs them.
      2. Make sure the use of alcohol is an adjunct to an activity rather than the primary focus of action.
      3. Make sure alcohol is not used in connection with other drugs.
      4. Make sure human dignity is served by the use of alcohol.
   B. When you are the host or hostess at a party:
      1. Have nonalcoholic drinks available.
      2. Do not pressure people to drink.
      3. Have plenty of food available.
      4. Serve coffee to allow time for sobering up.
      5. Don't let drunk guests drive home.

V. Young people and alcohol
   A. Young people are drinking more than ever before and in more damaging ways.
   B. The reasons are not clear.
   C. Efforts to prevent problem drinking in young people should be intensified.

VI. Problems resulting from alcohol abuse
   A. Alcohol and society—problem drinking may be both the cause of and result of social problems.
   B. Alcohol and the family—again, drinking can be both the cause of and result of family problems.
   C. Alcohol and crime—much crime is directly associated with drinking.
   D. Alcohol and driving
      1. Driving ability is impaired starting with the first drink.
      2. Drunk driving is predominantly a male behavior pattern.
      3. Drunk drivers are usually habitual problem drinkers.
   E. Alcohol and life span
      1. Moderate drinking does not shorten the life span.
      2. Heavy drinkers have a high death rate from many causes as well as a shortened life span.

7

ALCOHOLISM (ALCOHOL DEPENDENCE)

Alcoholism, compulsive abuse of alcohol, or alcohol dependence, is the number one hidden health problem in the United States. Current estimates show that some 9 or 10 million people are compulsive drinkers and that *everyone* is directly or indirectly affected by their alcohol abuse. (Those who try to estimate the incidence of alcoholism must use as a primary statistic the number of people who come to the attention of the police or who require medical attention for their alcoholism. This means that most estimates are probably on the low side.) Alcoholism unquestionably does extensive damage to individuals, to families, and to society as a whole. The alcoholic has physical health problems as well as difficulty holding a job, continuing his or her education, and maintaining a stable family life. Several forces act to keep much alcoholism hidden. A prominent factor is the unfortunate social stigma attached to alcoholism. Too many people mistakenly view alcoholism as a sign of some form of weakness, lack of "willpower," or some other personal inadequacy. Many are reluctant to admit to others and, more importantly, to themselves, that their drinking is having destructive effects on their lives. Furthermore, the treatment of alcoholism is not as simple and clear-cut as that of many other health problems. To many people, the available treatment methods seem nebulous, difficult to understand, tedious, and perhaps even threatening. As a result, many people choose to continue living with the problem of alcoholism; so consequently the disease remains hidden.

MEANINGS OF "ALCOHOLISM"

Various meanings have been given to the word "alcoholism" by different health professionals as well as the lay public. Some people suggest that it simply means "drinking too much." But what is "too much" drinking? Probably everyone has a different opinion. A more useful general definition might be "Alcoholism is the point at which a person's drinking interferes with some aspect of his or her life." This might include one's health, job, family life, or social interactions. An important point is suggested here: The way to judge if someone suffers from alcoholism is not to measure how much he or she drinks but to observe what effect drinking has on their life.

There are also restrictive definitions of alcoholism, such as the one which has been adopted by the National Council of Alcoholism:

> Alcoholism is a chronic, progressive and potentially fatal disease. It is characterized by tolerance and physical dependency, pathologic organ changes, or both, all of which are the direct or indirect consequences of the alcohol ingested.

According to this definition, the terms *chronic* and *progressive* imply that the physical, emotional, and social changes that develop are cumulative and progress as drinking continues. *Tolerance* means that the brain adapts to the presence of alcohol, requiring ever-increasing dosages in order to attain the same degree of effect. *Physical dependency* means that withdrawal symptoms may result from a decrease in the consumption of alcohol. Thus, this is a true addiction. The *pathological organ changes* referred to may be found in almost any organ but most often involve the liver, brain, nerves, and the digestive tract. The *social, emotional,* and *behavioral consequences* of alcoholism result from the effects of alcohol on brain function.

While not specifically stated in this definition, it is generally agreed that the drinking pattern of an alcoholic may be either continuous or intermittent with periods of abstinence. Of course, if drinking has progressed to the point of true addiction, drinking will be more or less continuous.

Finally, an outstanding characteristic of alcoholism is *loss of control*. This means the loss of the ability to drink in a moderate or controlled manner. Someone suffering from alcoholism cannot consistently predict on any drinking occasion the duration of the episode or the quantity of alcohol that will be consumed. The alcoholic person still retains control of whether or not he or she drinks on a particular occasion. But, if drinking does begin, it may be impossible to stop after a moderate amount. The person must continue to drink.

Some authorities classify alcoholism as either *primary* or *secondary* (Madsen, 1974). A primary alcoholic is invariably one who from early childhood has lived a life of loneliness, fed by a sense of inadequacy. For most primary alcoholics, the first exposure to alcohol is a "peak" experience (as described by A. H. Maslow, 1971, and McClelland, 1972). For the first time in their lives, they finally have a feeling of adequacy and a new sense of power that allows them to intcract freely with others. However, primary alcoholics rapidly lose their ability to drink in a controlled manner. A compulsion to drink increases, and within a few years, they are totally alienated, powerless, and confused.

Secondary alcoholism develops much more slowly. While primary alcoholics usually feel that they were "born alcoholic," secondary alcoholic individuals begin drinking in a controlled manner. The onset of this alcoholism may follow as much as 20 years or more of social drinking.

CAUSES OF ALCOHOLISM

Despite years of research efforts at a cost of millions of dollars, the causes of alcoholism are still not definitely known. Many theories have been presented, some of which are encouraged by extensive scientific evidence and some by pure speculation. It has not been clearly determined whether alcoholism is caused by physical factors, psychological factors, or a combination of the two. Each theory has strong supporters. There is certainly reason to believe that personality problems are a facet of alcoholism. Yet thcrc is also abundant evidence that some people simply respond differently to alcohol. This is probably related to metabolic differences. The question of whether alcoholism or the tendency toward alcoholism is hereditary is yet to be clearly resolved.

All authorities today do agree that alcoholism should be thought of as a disease, regardless of its cause, and that the alcoholic person should be treated as having an illness rather than condemned as a sinner or a good-for-nothing. If alcoholism were solely a problem of self-discipline, it would always be possible for the reformed alcoholic to return to controlled, social drinking. But this is almost never the case. A recovered alcoholic who has a single drink might well return to a completely alcoholic pattern of behavior, no matter how strong his or her resolve.

While there is no single "alcoholic personality," certain personality traits are commonly found among alcoholics. These traits are not exclusive to people with drinking problems, however, as they often occur in others as well. As a group, alcoholics are unhappy, have often suffered from deprivation of love and affection (particularly in childhood), and may have borne unusually heavy life stresses. They often experience difficulty in relating effectively to other people. Whether alone or with others, they feel a sense of loneliness and isolation from other people. Sometimes drinking and drinking friends seem to take the place of affectionate, meaningful relationships with others. (This may be a result of problem drinking, as well as a cause of it.) The need for immediate gratification is great; alcoholic people may have difficulty in handling the frustrations, delays, and anxieties of daily life.

Alcoholic people typically have low opinions of themselves. They often feel insecure and have little sense of personal worth. These feelings cause great emotional pain, and they may drink to wipe out this pain. These feelings are also common among other drug abusers and overeaters. As mentioned in Chapter 4, a "vicious circle" is frequently established, in which low self-esteem leads to substance abuse, which causes further loss of self-respect.

At least some research evidence has been presented to support each of the possible causes of alcoholism, but no single theory of causation has been adequately proven. In fact, there probably is no single cause of all alcoholism. More likely, alcoholism is the result of many complex factors acting together.

ALCOHOLISM AND WOMEN

Women have traditionally been "hidden" drinkers, often restricting their drinking to the home. This has made it difficult to estimate the numbers of alcoholic women and has also contributed to a tendency to ignore the problem as if it did not exist.

Despite a lack of hard statistics, authorities generally agree that the number of alcoholic women is somewhere between 2 and 4 million and is growing rapidly. The reasons for the increase are not entirely clear, but changing life-styles and values have undoubtedly had an effect, particularly as they have led to a redefining of roles for many women. Not only is the number of alcoholic women apparently increasing, but their drinking is more likely to present problems formerly associated with alcoholic men as the lives of women move beyond the confines of their homes.

Though there is some disagreement among experts, many feel that women begin to drink heavily later in life than do men, and then take less time to become alcoholic. Thus, the average age for women becoming alcoholic is about the same as that for men.

Women are the victims of a discriminatory social myth surrounding alcoholism. The nonalcoholic wife of an alcoholic man is frequently assumed to be contributing in some way to her husband's drinking. But the nonalcoholic husband of an alcoholic woman is more likely to be regarded as a deprived person, one who receives more sympathy than censure. Such attitudes don't help either sex in overcoming alcoholism and contribute to the high divorce rate of alcoholic people.

One particularly unfortunate aspect of alcoholism in women is its effect on any children they might bear. It has been shown (see *Alcohol and Health,* 1974, and *Medical Tribune,* March 16, 1977) that the incidence of birth defects is higher among babies born to alcoholic women. Serious malformations are commonly found in these babies, including abnormally small heads, poorly developed lower jaws, defects in bone and joint structure, and heart abnormalities. Furthermore, even if the baby is born without such problems, its chances for a fulfilling life are diminished. Alcoholism in either parent is a recognized factor in child abuse and child neglect.

Since alcoholic women have needs and characteristics that are somewhat different from those of alcoholic men, those treatment programs which have recognized these differences have been most successful in aiding the recovery of these women.

OLDER PROBLEM DRINKERS

Studies have shown that the incidence of drinking problems is lower among older persons in the United States than among younger people. The incidence of problem drinking among older people is apparently limited by several factors. For one thing, early death is much more common among long-term alcoholic people than among the general population. (See "Mortality of Alcoholic People," in *Alcohol Health and Research World,* Summer 1973.) Many alcoholic persons simply do not live long enough to be counted among the elderly. Furthermore, the current generation of older people in this country, even when younger, contained many more abstainers than is true of today's younger and "middle-aged" generations. Also, many people who are moderate-to-heavy users of alcohol in their youth decrease their drinking with increasing age, often for reasons of health.

Trend studies, however, indicate that the number of older problem drinkers is increasing as a higher number of people, especially women, are now entering old age with well-established drinking habits. Several factors have been identified as motivating excessive drinking among older persons. One is forced retirement, with its hours of boredom, loss of status, and lowered income. Many people are inadequately prepared for the adjustments necessary for a fulfilling life after retirement. Another factor is that older people have to cope with so many losses—loss of family, friends, status, health, and independence. Until our society comes to terms with the psychological and sociological factors of aging, it appears that the number of older problem drinkers will continue to increase.

PHASES OF ALCOHOLISM

Since at least 1 in every 20 North Americans eventually becomes alcoholic, it is important that anyone who uses alcoholic beverages be able to recognize the signs and symptoms of impending and early alcoholism.

ALCOHOLISM (ALCOHOL DEPENDENCE)

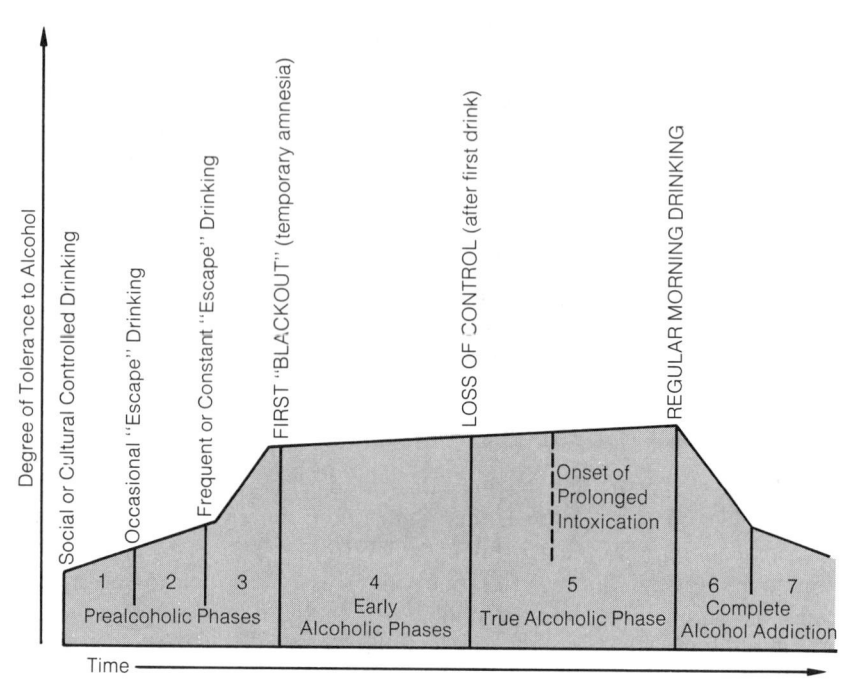

Figure 7.1. The phases of alcoholism.

Hopefully, if a person notices these symptoms developing and knows what they mean, he or she may have the good judgment to cease drinking before reaching a more advanced stage of alcoholism.

A classic Canadian research project in which the late Canadian researcher Dr. E. M. Jellinek interviewed over 2000 alcoholics revealed that most alcoholics pass through definite progressive stages with characteristic symptoms. This progress is shown in Figure 7.1. On this graph, time proceeds from left to right. There can be no fixed scale of months or years, because some alcoholics make the entire progression in a matter of months and others may take many years to reach the same point. Dr. Jellinek pointed out that the development of alcoholism in women is frequently more rapid than in men, but with the stages less clear-cut. The vertical line on the graph indicates the relative degree of alcohol tolerance, in other words, the amount of alcohol required to reach a given degree of intoxication.

Prealcoholic Phases

During the course of social drinking, most people learn the feeling of escape from everyday cares that alcohol can provide (phase 1 in Figure 7.1). An individual who occasionally drinks for the specific purpose of escape from tensions has progressed to the second prealcoholic phase. About 20 percent of all drinkers fall into this category. In the person destined for alcoholism, this escape drinking gradually becomes more and more frequent (phase 3).

As drinking becomes more frequent, a person develops an increased tolerance for alcohol; that is, he or she must drink more in order to achieve the same effect. At first, tolerance increases gradually; then it often takes a sudden jump. A high level of tolerance is maintained until, in late alcoholism, there is a marked loss of tolerance (phases 6 and 7).

Alcoholic Blackouts

Soon after this increase in tolerance, the frequent drinker may experience his or her first alcoholic blackout (phase 4). A blackout is a period of temporary amnesia occurring when a person is drinking. In contrast with passing out, which results in unconsciousness, a person in an alcoholic blackout is still conscious of what he or she is doing at the time and may be doing all the things that might normally be done. But after coming out of the blackout, there is no memory of anything that took place during the blackout. Anyone who drinks too much will pass out, but only the alcoholic or near-alcoholic blacks out. The mechanism of the blackout is not yet definitely known. It may have a physiological origin, or it may be a psychological ego-defense mechanism.

Loss of Control

Another major milestone in the development of alcoholism is loss of control (phase 5). This means the loss of the ability to drink in a moderate, controlled manner. It does not mean that the individual feels compelled to start drinking, but when drinking does start, it cannot stop after a predetermined, moderate amount of alcohol. A person at this stage may continue to drink until becoming quite intoxicated or sick. The period of drinking may last for a few hours or become a binge, lasting for days or weeks. The individual can choose whether to start drinking, but not when to stop it. By any definition of "alcoholism," this individual is now alcoholic.

Other Signs of Alcoholism

Many other symptoms are characteristic of the alcoholic person. These symptoms are arranged in sequence from the characteristics of early alcoholism to those of late alcoholism. Although, in a given individual, these symptoms may not occur in this exact order, they usually roughly follow this sequence.

ALCOHOLISM (ALCOHOL DEPENDENCE)

1. *Secret drinking.* The alcoholic person often sneaks drinks so that others will not know how much he or she is drinking.
2. *Preoccupation with alcohol.* The personal aspects of social functions become secondary to the opportunity for drinking. For example, when an alcoholic person is invited to a party, he or she may be more interested in the fact that drinks will be available than in who will be there.
3. *Gulping the first few drinks.* The alcoholic person drinks for the quickest possible effect.
4. *Guilt feelings about drinking.* As the people begin to realize that their drinking habits are not normal, they develop vague conscious or subconscious feelings of guilt that may lead to several outward symptoms.
    a. *Avoiding talk about alcohol.* People who eagerly talk about drinking are seldom problem drinkers. Alcoholic persons, in contrast, do not like to discuss drinking because they are afraid they may be criticized for their excessive drinking.
    b. *Rationalization of drinking behavior.* Alcoholic people always have a reason for drinking, which is actually an excuse or a rationalization. It never occurs to the normal drinker to offer a reason for drinking. For alcoholic people, though, good news and bad news are both valid reasons for drinking. They drink to celebrate their accomplishments or to drown their sorrows. These rationalizations are needed primarily for the protection of one's own ego and only secondarily as alibis for family and associates.
    c. *Exhibitions of grandiose behavior.* Alcoholic people often go through overly extravagant and generous periods during which they throw money around in a showy way. They may buy drinks for perfect strangers and leave unusually large tips. The purpose of such display is not so much to impress others as it is to reassure the alcoholic person that he or she is really not such a bad person after all. This is part of the system of rationalization that strongly influences the life of alcoholic people and serves to protect their ego.
    d. *Having periods of remorse.* Often the guilt feelings of alcoholic people lead to periods of persistent remorse that may have the unfortunate effect of leading them on to still more drinking.
5. *Periods of total abstinence.* As a result of social pressures or their own concern, alcoholic people may go "on the wagon." For several weeks or months, they may avoid taking a single drink. Then they usually resume drinking with renewed vigor because they are satisfied that they can still live without alcohol. Hence the oft-repeated reassurance, "I can take it or leave it," becomes the theme song of the alcoholic person.
6. *Changing drinking patterns.* Alcoholic people often feel that there

must be some way to drink without loss of control. In attempting to drink in a normal, controlled manner, they frequently vary their drinking patterns, trying different types of liquor, different mixers, or different times or places. Of course, none of these changes help.
7. *Behavior becomes alcohol-centered.* This symptom is characterized by a marked loss of interest in anything other than alcohol. Personal appearance is neglected, as is the maintenance of living quarters and possessions. There is a deterioration in interpersonal relationships. Instead of worrying about how drinking is affecting his or her activities, the alcoholic person may avoid activities that might interfere with drinking. The alcoholic person becomes increasingly egocentric.
8. *Effects on the family.* The members of the alcoholic person's family often change their habits. They may withdraw into the home for fear of embarrassment or, in contrast, may become very active in outside interests as a means of escape from the home environment. Financial problems are usually a way of life for the family of the alcoholic person.
9. *Unreasonable resentments.* Alcoholic people often build up tremendous feelings of resentment and self-pity. They may spend much time brooding over minor or imaginary injustices they have suffered.
10. *Hiding bottles.* The many jokes and cartoons about alcoholic people hiding (and losing) their bottles have a factual basis. The alcoholic person often takes elaborate precautions to avoid running out of liquor.
11. *Neglecting proper nutrition.* Chronic alcoholic people typically have little interest in food, deriving most of their calories from alcoholic beverages, which are very poor sources of vitamins, minerals, and proteins. They may suffer from serious malnutrition that may compound the physical damage resulting from the toxic effect of alcohol.
12. *Decrease in sexual drive.* As a result of a deteriorating physical and emotional condition, alcoholic people often suffer a decrease in sexual drive. This decrease often leads to alcoholic jealousy, in which the spouse of the alcoholic person is accused of having extramarital affairs. The marriage that has managed to survive to this point is often shattered by such jealousy.
13. *Regular morning drinking.* Alcohol is an addictive drug. After years of heavy drinking, a level of addiction may be reached that requires the constant presence of alcohol in the body in order to prevent withdrawal symptoms. This level of dependence indicates chronic alcohol addiction (phase 6 in Figure 7.1). The alcoholic person must now start each day with a drink (Figure 7.2). If the fully addicted alcoholic person is deprived of alcohol, the first withdrawal symptom is usually a shaking of the hands, arms, and body. The mood may be one of apprehension or fear, and hallucinations may occur. The alcoholic person at this stage should not be forced to

Figure 7.2. "*Open 6* A.M."

sober up without medical assistance because there is a chance of going into convulsions and perhaps even death.

14. *Intoxication during working hours.* Another result of the addiction aspect of alcoholism occurs when the alcoholic person is intoxicated in the morning on a working day and either misses work or sneaks drinks while on the job. This is often the beginning of the end of a job.
15. *Loss of tolerance.* After a long term of heavy drinking, alcoholic people lose the tolerance to alcohol they previously had acquired. This loss is possibly connected to a decrease in the ability of the liver to oxidize alcohol. Following this loss of tolerance, they become intoxicated on less liquor than before and "sober up" much more slowly. They are able to remain intoxicated at all times with only a moderate total consumption of alcohol.
16. *Mental impairment.* Many alcoholic people eventually suffer a severe disintegration of personality. If drinking continues after this symptom appears, there may be permanent brain damage. It is believed that this brain damage is due to the combined effect of severe dietary deficiency plus the direct effect of alcohol. A type of psychosis that commonly appears in late alcoholics is delirium tremens, the "DTs." This is a temporary condition, lasting only from two to

ten days, but there may be repeated attacks. An attack may be triggered by injury, illness, or withdrawal from alcohol. The symptoms may include confusion, vivid visual hallucinations, fear, apprehension, restlessness, and sleeplessness. There may even be convulsions similar to epileptic seizures. The attack ends with a period of deep sleep.
17. *Termination of alcoholism.* There are several possible ways in which a case of alcoholism may terminate. They include:
   a. *Failure of the rationalization systems.* Eventually, the elaborate system of rationalization that has sustained the alcoholic individual may break down, usually in response to some serious crisis. If the alcoholic person can admit defeat, he or she becomes readily accessible to treatment and can often successfully regain sobriety.
   b. *Wet brain.* After several attacks of delirium tremens, a more serious condition known as "wet brain" may develop. This is a chronic condition that is seldom curable and may even be fatal. The thought processes are completely disrupted, and all other functions of the nervous system are impaired. This is permanent brain damage, and the alcoholic person who reaches this stage either dies or spends the rest of his or her life in an institution.

Fortunately, it is not necessary to go through the entire course of the disease of alcoholism. An alcoholic person can be successfully treated at any stage if he or she is willing to admit to a drinking problem and really wants to do something about it.

True Alcohol Addiction (Phase 6)

True physical addiction usually develops only after years of heavy drinking. Prior to this development, drinking can be motivated by a complex of psychological, social, and physiological factors related to the individual and his or her environment. But once physical dependence develops, continued drinking may become a response to the threat of a withdrawal reaction should drinking be discontinued. In fact, someone at this late stage in the progression of alcoholism should discontinue drinking only with the aid of a physician. Drugs are given as a substitute for alcohol to prevent withdrawal symptoms. The patient is then gradually withdrawn from the substitute drug.

TREATMENT OF ALCOHOLISM

Because our knowledge of the causes of alcoholism is still fragmentary, it is reasonable to expect that controversies regarding its treatment might exist. This is certainly the case. For example, one major controversy that has emerged during the past few years is whether it is wise for alcoholic people to attempt to return to controlled drinking. The traditional wisdom has held that abstinence is the only effective way to deal with alcoholism. The majority of authorities on alcoholism are still convinced that, considering the very limited number of alcoholic persons who have successfully returned to controlled drinking, abstinence is the preferred goal in the treatment of alcoholism (see J. Doherty, 1974).

Initial Medical Treatment (Withdrawal)

Someone in the late stages of alcohol addiction should not just suddenly stop drinking, either voluntarily or involuntarily. When such a person is withdrawn from alcohol, extreme hypersensitivity to all external stimuli usually appears within a week after the blood-alcohol leaves have returned to zero. Such hypersensitivity in its most extreme form (delirium tremens or convulsions) requires emergency medical treatment. It is caused by the return to functioning of previously anesthetized neurons, aggravated by prolonged nutritional deficiencies.

The alcoholic person may require extended hospitalization at this time. Drugs such as tranquilizers, insulin, and thiamine are often given. Correction of dietary deficiencies is very important in the rehabilitation of alcoholic persons because their thought processes may be impaired by extreme malnutrition. In addition to being placed on a balanced diet these people are often given vitamin preparations.

Following the initial withdrawal or "drying out" phase of treatment, a second and longer-term of treatment must begin. This treatment usually includes exploring the forces that led to alcohol dependence, the role that alcohol has played in the life of the individual, and even the functions that the dependency itself has served. As mentioned in an earlier chapter, it is now recognized that in any substance dependency, the dependency or addiction itself comes to serve functions in the individual's life that go beyond the effects of the substance. For example, an individual's alcoholism may serve to keep that individual in a very comfortable state of dependency upon his or her spouse.

A few of the current approaches to the long-term treatment of alcohol abuse are discussed in the following section.

Drug Therapy

Many drugs have been tried in the treatment of alcoholic people, with varying success. The tranquilizing drugs are sometimes effective in decreasing such symptoms as anxiety, tremor, and restlessness, all common to alcohol addicts during the withdrawal phase.

An approach that has gained widespread popular interest is *aversion therapy*. This is the use of drugs that make a person sick if he or she drinks alcohol. This can be done in two ways. One is administering a daily dosage of a drug such as Antabuse (Disulfiram), which causes a highly unpleasant body reaction if any alcohol is then consumed. Breathing becomes difficult; the heart pounds; and nausea and vomiting may occur. As long as a person is taking Antabuse, he or she is not likely to drink. This drug may be taken for months or even years, but the person must really want to stop drinking. Otherwise, he or she will just stop taking the Antabuse.

Another type of aversion therapy is to give a person a drink of liquor along with a drug that causes immediate nausea. After several of these treatments, a conditioned reflex may develop so that alcohol alone will cause illness.

The problem with either type of aversion therapy is that severe psychotic symptoms may occur if the alcoholic individual is suddenly deprived of escape. Most problem drinkers have become dependent on alcohol as an escape from life; if no effective psychotherapy is given, they may undergo severe emotional stress and personality disintegration without alcohol.

Psychotherapy

The success of psychotherapy in treating drinking problems has been variable. As in other areas of psychotherapy, the approach may be either a superficial emotional reinforcement of the individual or a deep exploration of the subconscious mind to uncover underlying emotional conflicts that may be contributing to the alcoholism. For success, the therapist must have an unusually good understanding of the alcoholic personality. It is very difficult for someone who has never had a drinking problem to understand what it means to have one. Group psychotherapy may be especially helpful because a group of alcoholic people will understand each other.

Alcoholism-Treatment Facilities

Many of the therapeutic communities (discussed in Chapter 5 in relation to the treatment of drug abuse) also admit people with drinking problems. The founders of these communities were among the very first to recognize alcoholism as a drug addiction. Synanon, in California, was the first to open its membership to alcoholic people, and a large number of its membership have had drinking problems.

Other than these, there are still very few effective facilities for the continuing treatment of alcoholic people. There are a few private or government outpatient clinics and city or county hospitals willing to treat alcoholism. Many large corporations now have special educational and treatment programs to help those of their employees who have drinking problems.

Alcoholics Anonymous

One of the most successful approaches to the treatment of alcoholism has been that of Alcoholics Anonymous (commonly called AA). It is believed that AA has the greatest recovery rate (75 percent of those who want to stop drinking) of all methods of treatment of alcoholism.

Alcoholics Anonymous is an organization whose only purpose is to help its members stay sober. Today, almost every city has regularly meeting AA groups ranging in size from a handful of members to over a hundred. A large city might have groups meeting every night of the week. There are even special groups for teen-age alcoholics and for spouses and children of alcoholics.

The approach taken by AA is that of group therapy. Like the "dope fiends" of Synanon groups, which to some extent are patterned after AA meetings, alcoholic people often find a deep personal, emotional, and spiritual experience through close association and conversation with others who have shared their addiction. An evening's program usually consists of several members telling informally how miserable their lives were during their drinking years and how they have changed since joining AA. New members often find that these admitted alcoholics have had experiences similar to their own. They can identify with the older members, who "speak the same language." As older members tell of their past experiences, they are also helping themselves to stay sober. The stories serve as a constant reminder of the unhappiness of their periods of drinking; they help to prevent them from returning to drinking.

Alcoholics Anonymous does not claim to cure alcoholic people; rather, it helps them to stop drinking and regain sobriety. It emphasizes that alcoholic drinkers remain vulnerable to alcohol addiction. If they start drinking again, they will still drink in an alcoholic manner. For this reason, members always begin their personal stories by stating "I am an alcoholic."

Also, it must be stressed that AA cannot help people who do not have three essential qualifications: (1) a sincere desire to stop drinking; (2) a willingness to admit that they, by themselves, are unable to solve their drinking problem and, therefore, that they must have help; and (3) the ability to be honest with themselves.

There have been many cases where members of Alcoholics Anonymous decided, after years of sobriety, to try a return to social drinking. These attempts are very rarely successful. Alcoholics Anonymous can only help those who have a strong desire to stop drinking forever. This is why AA is strongly opposed to any effort to return alcoholic people to controlled drinking.

SUMMARY

I. Meanings of the word "alcoholism"
   A. A general definition: Alcoholism is the point at which a person's drinking interferes with some aspect of his or her life.
   B. A specific definition: alcoholism is a chronic, progressive and potentially fatal disease, characterized by tolerance and physical dependency, pathologic organ changes, or both, all of which are the direct or indirect consequences of the alcohol ingested.
   C. Some authorities classify alcoholism as primary or secondary:
      1. Primary—alcoholism that develops very rapidly after a person begins drinking.
      2. Secondary—alcoholism that takes years to develop.

II. Causes of alcoholism
   A. The causes are still not definitely known, although many theories exist.
   B. It may be caused by physical or psychological factors or a combination of the two.
   C. Alcoholism should be thought of as a disease, not a sin or a weakness.
   D. While no single "alcoholic personality" exists, certain personality traits are common among alcoholic people.
      1. Unhappy.
      2. Difficulty in relating to other people.
      3. Need for immediate gratification.
      4. Low self-esteem.

III. Alcoholism and women
   A. Women have traditionally been "hidden" drinkers.
   B. The number of alcoholic women is now growing rapidly.
   C. Babies of alcoholic women have increased incidence of birth defects.

IV. Older problem drinkers
   A. The incidence of drinking problems has been lower in older persons than in younger people.
   B. The current trend is increasing incidence of alcoholism in older people.

V. Phases of alcoholism
   A. Many alcoholics pass through definite progressive stages in their disease.

ALCOHOLISM (ALCOHOL DEPENDENCE)

- B. Prealcoholic phases
    1. Social drinking.
    2. Occasional escape drinking.
    3. Frequent escape drinking.
- C. Alcoholic blackouts—temporary amnesia.
- D. Loss of control—loss of ability to drink in a controlled manner.
- E. Other signs of alcoholism (see text).
- F. True alcohol addiction—the need to drink to avoid physical withdrawal illness.

VI. Treatment of alcoholism
- A. Many controversies exist.
- B. Initial medical treatment (withdrawal)—requires the aid of a physician for the fully addicted alcoholic.
- C. Drug therapy (aversion therapy)—drugs make the person sick if he or she drinks alcohol.
- D. Psychotherapy—variable success.
- E. Alcohol-treatment facilities—more are needed.
- F. Alcoholics Anonymous (AA)
    1. It has been one of the most successful approaches.
    2. It provides a form of group therapy.
    3. It does not attempt to return alcoholic persons to controlled drinking—the emphasis is on abstinence.

GLOSSARY

For definitions of terms not included here, consult the Index for text references.

abstinence syndrome: Set of symptoms resulting from withdrawal from alcohol, opiates, and specific depressants

abuse: To use wrongly or improperly; misuse

acetylcholine: Chemical liberated at the synapses of certain nerve cells; responsible for transmission of the nerve impulses

acetylsalicylic acid: Aspirin; used as antirheumatic, analgesic, and antipyretic

addiction: Condition resulting from repeated use of a drug in which physical dependence is established because of biochemical and physiological adaptations to the drug

additive effect: Sum of the effects of two or more drugs that produce similar effects when such drugs are administered in combination

alcohol: Chemical name *ethanol*. A transparent, colorless, volatile liquid obtained by fermentation of carbohydrates with yeast

amine: Chemical group NH_2; as a prefix (amino-), indicates the presence in a compound of the group NH_2

amnesia: Loss of memory or loss of a large block of interrelated memories

amphetamine: Drug group that acts as a stimulant to the central nervous system

analgesic: Chemical or drug that has the ability to relieve pain

GLOSSARY

anesthetic: Chemical, substance, or agent that produces loss of feeling or sensation.

antacid: Substance that counteracts or neutralizes acidity, usually of the stomach

antagonistic: Opposing or counteracting drugs or medicines

antibiotic: Chemical substance produced by a microorganism that has the capacity to inhibit the growth of or to kill other microorganisms

antidepressant: Drug that counteracts the feelings of depression (absence of cheerfullness or hope) with reduction of the functional activity of the body

antiemetic: Drug, medicine, or substance that prevents vomiting or relieves nausea

antihistamine: Drug that counteracts the action of histamine

aphrodisiac: Any drug that arouses sexual excitement

ascites: Accumulation of fluid in the abdominal cavity; also called *abdominal* or *peritoneal dropsy*

autonomic nervous system: Branches of cranial and spinal nerves that supply the visceral organs

aversion (therapy): Therapy associating an undesirable behavior pattern with unpleasant stimulation or making the unpleasant stimulation a consequence of the undesirable behavior

barbiturate: Drug group used in medicine as hypnotics or sedatives

Benzedrine: Trade name for a specific amphetamine

biochemical: Pertaining to the chemistry of living matter; chemistry of the body

blackout (alcohol induced): When used in reference to alcohol abuse, a period of temporary amnesia occurring while drinking alcoholic beverages

caffeine: Alkaloid chemical found in coffee, tea, and other substances; a stimulant to the central nervous system

cardiovascular: Pertaining to the heart and blood vessels

catecholamine: Group of compounds having a sympathomimetic action. Such compounds include dopamine, norepinephrine, and epinephrine

cathartic: Agent that causes evacuation of the bowels by increasing bulk or stimulating peristaltic action

chemotherapy: Treatment of disease by the administration of chemicals, medicines, or drugs

chloral hydrate: Crystalline substance with an aromatic, penetrating odor and a bitter, caustic taste; used as a hypnotic

chronic: Continuing for a long time; more specifically, used to describe condition of body or disease that is of long duration

cirrhosis: Disease of the liver involving progressive destruction of liver

GLOSSARY

cells, accompanied by excessive increase in connective tissue, resulting in contraction of the organ
cocaine: Drug used in medicine as a local anesthetic; a stimulant to the central nervous system
codeine: Analgesic, hypnotic-sedative drug derived from opium; classified as a narcotic
coma: Abnormal, deep stupor or sleep
comatose: In a coma
conditioned reflex: Learned response to a specific stimulus that has been repeated until it appears automatic
congener: Chemical compound closely related to another in composition and exerting similar or antagonistic effects
convulsions: Contortions of the body caused by violent involuntary muscular contractions

dangerous drugs (legal term): Drugs, other than narcotics, controlled by law
delirium tremens: Psychic disorder involving visual and auditory hallucinations, delusions, incoherence, anxiety, and trembling; found in habitual (addicted) users of alcoholic beverages and some other drugs
denature: To change the nature of a substance; in reference to alcohol, to make it unfit for human consumption without affecting its usefulness for other purposes
dependency: Being controlled or influenced by something else
depersonalization: Loss of the sense of personal identity or of personal ownership of the parts of one's body
depressant: Chemical, substance, or agent that has the ability to reduce the functional activity of the body
depression: Reduction of the functional activity of the body
Dexedrine: Trade name for a specific amphetamine
distillation: Process of vaporizing a liquid and then condensing the products of vaporization into a liquid in order to purify or abstract a specific substance from the original liquid
dopamine: Intermediate product in the synthesis of norepinephrine. Also, see *catecholamine*
dosage: The amount of a drug to be given, or taken, at one time
drug dependence: Condition resulting from repeated use of a drug in which individual must continue to take drug to avoid abstinence syndrome (physical dependence) or to satisfy strong emotional need (psychic dependence)
dysfunction: Abnormal, impaired, or incomplete functioning of an organ or part
dysorganization: State of impaired and inefficient emotional organization

resulting from a person's inability to cope with internal conflicts and external reality

egocentric: Regarding self as the center of all things

ego (processes): According to Freudian theory, one of the three major divisions of the psychic apparatus; consciously acts as mediator for the impulses of the id, the prohibitions of the superego, and the demands of reality

endocrine: Any of the glands of internal secretion that produce hormones; also, the secretion produced by any of these glands

epinephrine: Monoamine hormone secreted by the adrenal medulla; responsible for transmission of nerve impulses to organs of the sympathetic nervous system

euphoria: Feeling of well-being; in psychiatry, exhibiting an abnormal or exaggerated sense of well-being

felony: Offense punishable by death or imprisonment for more than one year

gastritis: Inflammation of the stomach

generic: Nonproprietary; denoting a drug name not protected by a trademark; sometimes called *public name*

gratification: Source of pleasure or satisfaction

habituation: Condition resulting from repeated use of a drug in which a psychic, not a physical, dependence is established

hallucination: Perception of objects or experience of sensations with no real external cause

hallucinogen(ic): Chemical, substance, or agent capable of producing distortions of the senses that may include hallucinations

hedonistic: Pertaining to pleasure

heroin: Diacetylmorphine. A white, bitterish, crystalline powder; an analgesic and narcotic

hydrocarbon: Class of chemical compounds containing only hydrogen and carbon

hyperalert: Abnormally and excessively watchful of activities around oneself

hypnotic: Drug that acts to induce sleep

id (processes): According to Freudian theory, one of the three main divisions of the psychic apparatus; harbors the unconscious instinctive desires and striving of the person

illicit: Unlawful; not allowed by law or custom

impotence: Lack of ability to engage in sexual intercourse

intoxication: Literally, state of being poisoned or drugged; condition

produced by excessive use (abuse) of toxic drugs, alcohol, barbiturates, and so forth
intracellular: Within a cell or cells
intravenous: Into a vein
irritability: Ability to respond to stimuli

lactose: Sugar found in milk
latency: Period of time between the administration of a drug and the beginning of response
length of action: Period of time when a drug is effective
lethal: Causing death
leukopenia: Reduction of the number of white blood cells (leukocytes) in the blood

malnutrition: Lack of proper nutrition; lack of adequate food or proper food
mania: Phase of mental disorder characterized by an expansive emotional state, elation, hyperirritability, overtalkativeness, flight of ideas, and increased motor activity
marijuana: Mexican name, of doubtful origin, for the plant *Cannabis sativa* or parts of it
medicinal: Having healing qualities
mescaline: Toxic, hallucinogenic oil extracted from peyote (*Lophophora williamsii*)
meprobamate: White powder, soluble in water; used as a minor tranquilizer; sold under trade names Miltown and Equanil
metabolite: Any substance produced by metabolism (the living functions of the cells and the body)
methadone: Synthetic narcotic similar to morphine and heroin; used as an analgesic and as a substitute narcotic in treatment of heroin addiction
misdemeanor: Offense defined as less serious than a felony; punishable by less than one year in jail
monoamine: All chemicals having a benzene ring and one amino nitrogen atom separated by a chain of carbon atoms (see Figure 3.17)
monoamine oxidase: Enzyme that catalyzes monoamines
morphine: Widely used analgesic and sedative; classified as an opiate narcotic
motor: Muscle, nerve, or brain center that affects or produces movement in the body
mucous membrane: Membrane that secretes mucus; the internal linings of the body

Nalline: Trade name for *nalorphine*. White odorless powder; used as an antidote to narcotic overdosage

GLOSSARY

narcotic: Having the power to produce a state of sleep or drowsiness and to relieve pain

narcotic antagonist: Substance that opposes the action of narcotics on the nervous system. See also *antagonist*

neuromuscular junction: Junction between nerves and muscles

neurotransmitter (transmitter or excitatory transmitter): Chemical substance that induces activity in a nerve or muscle

nicotine: Toxic alkaloid drug that is the active ingredient of tobacco

norepinephrine: Neurohormone responsible for transmission of nerve impulses from the ergotropic division of the brain to the sympathetic nervous system

nystagmus: Involuntary rapid movement of the eyeball

opiate: Any drug containing or derived from opium; true narcotics

opium: Narcotic drug derived from the dried juice of the opium poppy *(Papaver somniferum)*

oral: Related to the mouth or swallowing

organotropic: Having an attraction for certain organs or tissues of the body

oxidation: Combination with oxygen or removal of hydrogen from a compound by the action of oxygen; one means by which body disposes of foreign substances such as alcohol

paraldehyde: Hypnotic drug derived from alcohol

parasympathetic nervous system: Branch of the autonomic nervous system that consists of a group of cranial nerves that leave the spinal column in the sacral area of the body; generally causing a relaxation of the body and automatic functions of the body

patent medicine: A drug or medicine's exclusive right of production, use, or sale is owned by an individual or company

pathological: Diseases; especially diseased structural and functional changes in tissues and organs of the body

pentobarbital: Short-acting barbiturate

permeability: Ability to allow passage, as through a cell membrane

personality: Total reaction of a person to his or her environment

peyote: Common name for the cactus *Lophophora williamsii*

physiological: Pertaining to the functioning of the body

pons: Portion of the brain lying between the medulla oblongata and the mesencephalon, beneath the cerebellum

potentiation: Combined action of two drugs that is greater than the sum of the effects of each used alone

proof (of an alcoholic beverage): The relative strength of an alcoholic beverage with the reference standard for proof taken as 100 proof. One hundred proof is equal to 50 percent alcohol

proprioceptive: Receiving stimulations from the tissues of the body

GLOSSARY

proprioceptor: Sensory nerve terminal that gives information concerning movements and position of the body

psychic energizer: Drug group that acts as a stimulant to the central nervous system and is able to produce an elevation of mood, increased activity, heightened confidence, and an increased ability to concentrate

psychoactive: Exerting an effect upon the mind or psyche; capable of modifying mental activity; term usually applied to drugs that affect the mental state

psychogenic: Originating in the mind

psychotherapy: Treatment of psychological abnormalities or disorders

psychotomimetic: Producing manifestations resembling those of a psychosis; term applied to drugs that produce these effects

psychotoxic: Literally, poisonous to the mind; having the ability to modify mood and change behavior

rationalization: Invention of an acceptable explanation for behavior that has its true origin in the unconscious

receptor: Specific chemical grouping on the surface of a cell with the capability of combining with another specific chemical

regression: Return to a previous state of emotional development

replacement: Supplement or restorative; chemicals or drugs used to restore or supplement some secretion of the body

sedative: Depressant drug that allays excitement; quieting

self-actualization: The realization of one's full potential as a human being

semicoherent: Being partially intelligible, logical, and able to articulate

serotonin: Neurohormonelike substance found in many cells and organs of the body; responsible for activation of the trophotropic system in the brain, transmitting nervous impulses to the parasympathetic nervous system

serum: Clear, liquid portion of an animal fluid separated from the solid elements; often, blood serum

somnolence: Sleepiness; unnatural drowsiness

stimulant: Chemical, substance, or agent that has the ability to increase the functional activity of the body

strychnine: Alkaloid from seeds of *Strychnos nux-vomica,* an extremely potent stimulant to the central nervous system

subcutaneous: Under the skin

superego (processes): According to Freudian theory, one of the three main divisions of the psychic apparatus; associated with ethics, standards, and self-criticism which help form the conscience

sympathetic nervous system: Branch of the autonomic nervous system consisting of a pair of long nerve trunks located on either side of the backbone that supply the visceral organs; generally causing an excited or ready reaction in the body

sympathomimetic: Mimicking the effects of impulses affecting the sympathetic nervous system; drugs producing effects similar to stimulation of the sympathetic nervous system

symptomatic: Pertaining to or of the nature of a symptom

syndrome: Set of symptoms which occur together; the sum of signs and symptoms of a diseased state

synergism: Effect of a combination of drugs which, when taken simultaneously, have an effect greater than the sum of the same drugs when taken separately

synthetic: Formed by a chemical reaction in a laboratory

systemic: Affecting the body as a whole

therapeutic: Ability to heal

tranquilizer: Agent that acts on the emotional state, quieting or calming someone without affecting clarity of consciousness

thrombocytopenia: An abnormal increase in blood platelets (circular or oval disks, found in the blood, important in the clotting of blood)

tincture: An alcoholic solution of a chemical substance

tolerance: Increasing resistance to the usual effects of a drug

toxicity: Quality of being poisonous

tranquilizer: Drug that acts to relieve an overactive, anxious, or disturbed emotional state; a central nervous system depressant

untoward: Inconvenient, unfortunate, or unfavorable

vaccine: Preparation of killed or weakened microorganisms for use in immunization

vasodilator: Causing the blood vessels to dilate (expand in size)

vector: Carrier, especially an animal (usually an arthropod), which transfers an infective agent from one person to another

volatile solvent: Easily vaporized substance capable of dissolving something; specifically, chemicals contained in lighter fluid, model airplane glue, and other common substances that produce a state of intoxication when inhaled

withdrawal: Abstention from drugs to which one is habituated or addicted; also, term denoting the symptoms of such withdrawal

BIBLIOGRAPHY

"Alcoholism and Women." *Alcohol Health and Research World* (Summer 1974): 2–7.

Baker, Stewart, L., Jr. "U.S. Army Heroin Abuse Identification Program in Vietnam: Implications for a Methadone Program." *American Journal of Public Health* 62, no. 6 (June 1972): 857–860.

Bellwood, Lester. "Grief Work in Alcoholism Treatment." *Alcohol Health and Research World* (Spring 1975): 8–11.

Bergersen, Betty S., and Krug, Elsie E. *Pharmacology in Nursing.* 11th ed. St. Louis: Mosby, 1971.

Birdwood, George. *Willing Victim: A Parent's Guide to Drug Abuse.* New York: International Publishers, 1970.

Brenner, Joseph H., Coles, Robert, and Meagher, Dermot. *Drugs and Youth.* New York: Liveright, 1970.

Chafetz, Morris E., *et. al. Alcohol and Health: New Knowledge.* Rockville, MD: National Institute on Alcohol Abuse and Alcoholism, 1974.

Conn, Howard, ed. *Current Therapy, 1978.* Philadelphia: Saunders, 1978.

Doherty, James. "Controlled Drinking: Valid Approach or Deadly Snare?" *Alcohol Health and Research World* (Fall 1974): 2–8.

———. "Disulfiram (Antabuse): Chemical Commitment to Abstinence." *Alcohol Health and Research World* (Spring 1976): 2–8.

"Exploring the Nature of Heroin Addiction." *Medical World News* 12, no. 33 (10 September 1971): 57–63, 66.

Fallding, H. and Miles, C. *Drinking, Community and Civilization: The Account of a New Jersey Interview Study.* Newark, N.J.: Rutgers Center for Alcoholism, 1974.

Garb, Soloman, Crim, Betty Jean, and Thomas Garf. *Pharmacology and Patient Care.* New York: Springer, 1970.

Garmon, William S., and Strickland, Phil. *How to Fight the Drug Menace.* Boston: MIT Press, 1971.

Gibbins, Robert J., et al. *Research Advances in Alcohol and Drug Problems.* vols. 1 and 2. New York: Wiley, 1974.

Goldstein, A., Arnonow, Lewis, and Kalman, Summer M. *Principles of Drug Action: The Basis of Pharmacology,* 2nd ed. New York: Wiley, 1974.

Goodman, Louis S., and Gilman, Alfred, eds. *The Pharmacological Basis of Therapeutics.* 4th ed. New York: Macmillan, 1970.

Hindman, Margaret. "Family Therapy in Alcoholism." *Alcohol Health and Research* (Fall 1976): 2–9.

Hoff, Ebbe Curtis. *Alcoholism: The Hidden Addiction.* New York: Seabury Press, 1974.

Huey, Florence L. "In a therapeutic community." *Am. J. Nursing* 71, no. 5 (May 1971): 926–933.

Jaffee, Jerome H., and Senay, Edward C. "Methadone and methadly acetate: Use in management of narcotics addicts." *Mental Health Digest* 3, no. 8 (August 1971): 38.

Jellinek, E. M. *The Disease Concept of Alcoholism.* Highland Park, N.J.: Hill-House, 1960.

———. "Phases of alcohol addiction." *Quart. J. Studies on Alcohol* 13 (1952): 673–678.

Jones, Ben Morgan and Jones, Marilyn K. "States of Consciousness and Alcohol." *Alcohol Health and Research World* (Fall 1976): 10–15.

Krystal, H., and Raskin, Herbert A. *Drug Dependence Aspects of Ego Functions.* Detroit: Wayne State University Press, 1970.

Leavit, Fred. *Drugs and Behavior.* Philadelphia: Saunders, 1974.

Lieber, Charles. "Alcohol and the Liver." *Alcohol Health and Research World* (Spring 1974): 23–26.

Lieber, Charles S. "The Metabolism of Alcohol." *Scientific American* 234, no. 3 (March 1976): 25–33.

Madsen, William. "Alcoholics Anonymous as a Crisis Cult." *Alcohol Health and Research World* (Spring 1974): 27–30.

Martin, Eric W. and Martin, Ruth D. *Hazards of Medication.* Philadelphia: Lippincott, 1970.

Marx, Jean L. "Neurobiology: Researchers High on Endogenous Opiates." *Science* 193, no. 24 (September 1976): 1227–1229.

Maslow, A. H. *The Farther Reaches of Human Nature.* New York: Viking Press, 1971.

Maxwell, Kenneth E. "Mind Modifiers." *Chemicals and Life.* Belmont, Calif.: Dickenson, 1970.

McClelland, D. C., et al. *The Drinking Man.* New York: The Free Press, 1972.

Nowlis, Helen. *Drugs on the College Campus.* New York: McGraw-Hill, 1969.

———. "Perspectives on drug use." *J. Social Issues* 27, no. 3 (1971): 7–21.

"Older Problem Drinkers." *Alcohol Health and Research World* (Spring 1975): 12–17.

Ottenberg, Donald. "Addiction as Metaphor." *Alcohol Health and Research World* (Fall 1974): 18–20.

Physicians Desk Reference. 32nd ed. Oradell, N.J.: Medical Economics Co., 1978.

"Psychodrama in Alcoholism Treatment." *Alcohol Health and Research World* (Summer 1975): 11–13.

Puig, Margarite M., Gascon, Pedro, Cravisco, Gale L., and Musacchio, Jose M. "Endogenous Opiate Receptor Ligand: Electrically Induced Release in the Guinea Pig Ileum." *Science* 195, no. 28 (January 1977): 419–420.

Rabkin, Judith G. and Struening, Elmer L. "Life Events, Stress, and Illness." *Science* 194, no. 3 (December 1976): 1013–1020.

Regan, Timothy. "Cardiac Toxicity of Ethyl Alcohol." *Alcohol Health and Research World* (Winter 1973/74): 23–28.

Room, Robin and Sheffield, Susan, eds. *The Prevention of Alcohol Problems (Report of a Conference).* Berkeley: The University of California Press, 1974.

Robinson, Arthur L. "Analgesia: How the Body Inhibits Pain Perception." *Science* 195, no. 4 (February 1977): 471–474.

Rosenbaum, Max. *Drug Abuse and Drug Addiction,* rev. ed. New York: Gordon and Breach, 1974.

Shafer, Raymond P., Chairman. *Marihuana: A Signal of Misunderstanding.* Washington, D.C.: U.S. Government Printing Office, 1972.
State of California, Department of Justice. *Uniform Controlled Substances Act, 1976.* Sacramento, California: Bureau of Narcotic Enforcement, 1976.

Watson, Robert. "Alcohol: Denatured and Illegal Varieties." *Alcohol Health and Research World* (Spring 1975): 18–19.
Wei, Eddie, and Loh, Horace. "Physical Dependence on Opiate-like Peptides." *Science* 193, no. 2 (February 1976): 1262–1263.
Wolfgang Schmidt, Jur, and de Lint, Jan. "The Mortality of Alcoholic People." *Alcohol Health and Research World* (Summer 1973): 16–20.

Young, Alex W. 'Cutaneous Stigmata of Alcoholism." *Alcohol Health and Research World* (Summer 1974): 24–28.
"Young People and Alcohol." *Alcohol Health and Research World* (Summer 1975): 2–10.

Zarofonetis, C. J., ed. *Drug Abuse: Proceedings of the International Conference.* Philadelphia: Lea & Febiger, 1972.

INDEX

Abusive dosage, 47
Abstinence syndrome, 119
Acetaldehyde, 175
Acetylcholine (Ach), 28, 43, 44
Acetylsalicylic acid, *See* Aspirin
Acupuncture, to treat narcotic adliction, 61
Addiction, 117–119
 levels of, 51
Adrenalin. *See* Epinephrine
Adrenergic, 29–30
 drugs, 44
 fibers, 44
Adultrate, 68
Alcohol, 9–12, 14, 15, 51, 77–78
 alcohol-sedative withdrawal syndrome, 77
 absorption of, 167
 effects of, 166–175
 effects on babies, 186–187
 states of consciousness, 171–72
 and society, 177–178
 types of, 163
Alcoholic, 119
Alcoholic beverages, 12–13
 nutritional value of, 164–165
Alcoholics Anonymous (AA), 196–197
Alcoholism, 183–197
 causes of, 185–186
 definitions of, 184
 phases of, 187–193
 primary, 184–185
 secondary, 184–185
 signs of, 190–193
 treatment of, 193–197
Alert, 36
Alertness, 39
Amine theory of mood, 98
Amphetamines, 10, 12, 13, 15, 16, 18, 101
Amygdaloid nucleus, 42
Amyl nitrite, 72
Analgesia, 59
Androgens, 14, 17
Anesthesia, 59

Anesthetics, 9, 12, 14, 57–58
 local, 57
 general, 57–58
Angel dust, 57
Antabuse, 195
Antacids, 10, 11, 14, 17
Antagonist, 31
Antianxiety drugs, 84
Antibiotics, 11, 56
Anticholinergics, 101
Anticoagulants, 9, 11, 12, 14–17
Antidepressants, 9, 11–14, 18, 98–101
Antidiabetic drugs, 9, 13, 15, 17
Antihistamines, 10, 12, 17–18, 57, 79–81
Antilypertensive drugs, 12–13
Antiinflammatory drugs, 13
Antipsychotic agents, 82
Appetite, 38
Arousal, reaction, 39–40
 center, 40
Aspirin, 7, 9, 11–14, 17, 56
Atarax, 84
Atropine, 101
Autonomic nervous system, 44

Barbiturates, 12, 14–15, 17, 51–52, 57, 73–77
Basic human needs, 124
Behavior
 levels of, 44
Benadryl, 79
Benzedrine, 103, 106
Bile, 33
Black Widow spider (venom), 15
Blackout, alcoholic, 189
Blood
 alcohol levels, 168
 cholesterol, 15
 pressure, 38
Boggs Act, 139
Bromides, 78
Bufotenine, 93

211

INDEX

Bureau of Narcotics, 136
Butyl nitrite, 72

Caffeine, 15, 103
Cannabis, 84–89
Cap, 68
Cardiac center, 38
Cardiopulmonary resuscitation (CPR), 64–65
Catecholamines, 29
Cathartics, 16
Catnip, 56
Central nervous system (CNS), 28
  depressants, 9, 12, 14
  stimulants, 9
Central gray area, 59
Cerebrum, 34, 37, 38
Cholinergic
  drugs, 44
  fibers, 28, 43
Chloral hydrate, 77–78
Chloroform, 57
Cingulate gyrus, 42
Cirrhosis, 173
Cocaine, 10, 13, 15, 96–98
Codiene, 69
Cold remedies, 15, 56
Coma, 34
Commitment
  civil, 149
  criminal, 149
Communities, therapeutic, 154–156
Comprehensive Drug Abuse Prevention and Control Act, 141
Congeners, 164, 175
Conscious sensations, 34
Continuum of drug actions, 81
Cope, 81
Cough
  center, 39
  remedies, 56
Cox, B. M., 59
CPR. *See* Cardiopulmonary resuscitation
Crime, effects of alcohol on, 178–179
Cultural adjustment, 62
Cut, 68
Cyclamates, 11
Cyclazocine, 153

Darvon, 15
Daytop Village, 155
Death, 34
Delirium tremens, 119
Dependency, 117–122
Depressants, 34, 51
Dexedrine, 103, 106
Diacetylmorphine. *See* Heroin
Dietary supplements, 15–16
Digestive system, effects of alcohol, 172–173
Digitalis, 11, 16
Disulfiram, 10
Dizziness. *See* Equilibrium
DMT, 92
Docility, 46
DOM, 92
Dopamine, 29
Doped, 62
Doriden, 79
Downers, 56
Dramamine, 79
Driving, effects of alcohol, 179
Drowsiness, 34, 45
Drug Abuse Control Amendments, 140
Drug Enforcement Administration, 135, 140
Drugs, 1
  actions, 22–24

dosages, 20–21
elimination, 31–33
idiosyncrasy, 97
induced effects, 33–34
interactions, 24
maintenance therapy, 154
metabolism, 31
users, 2
Durham-Humphrey Act (Amendment), 6, 139

Earle, Robert W., 47
Ego, 123, 126
Electrostimulation, for pain, 61
Emergency therapy, 152
Emotional dysorganization, 126–129
Emotions
  alterations of, 33
  levels of, 44
Endocrine system, 38
Endorphin, 60–62
Enkephalin, 60–62
Enzyme, 31, 46
Epinephrine, 29
Equanil, 84
Equilibrium, 38
Ergotropic division, 44–45
Ethchlorvynol, 79
Ether, 57
Ethinamate, 79
Euphoria, 31, 62
Experimenters, 116

Family, effects of alcohol, 178
Fatty liver, 173
Federal Food, Drug and Cosmetic Act, 6, 139
Fluid balance, 38
Fornix, 42
FRAT test, 151
Free radical assay technique, 151
Freud, Sigmund, 123
Frozen pupils, 62

Ganglion, 43
Gateway House, 155
Glutethimide, 79
Goldstein, Auram, 59

Habituation, 117–118
Hallucinations, 50
Hallucinogens, 90–97
Hangover, 164, 175
Harrison Narcotics Act, 138
Hashish, 84–85
Heart, effects of alcohol, 173–174
Heaving, 34
Herbs, 56
Heroin, 15, 68–69
  ingestion of, 69
  tolerance to, 69
Hippocampel gyrus, 42
Hippocampus, 42
Histamine, 79
Hormones, 38, 59–60
Hughes, John, 50
Hyperkinetic, 104
Hypnotics, 13, 51
Hypnotic sedative drugs, 73–81
Hypothalamus, 38, 44, 46

Id, 123
Imagination, 36
Incus, 42
Indole alkaloids, 92
Inhalation, 3
Injection, 3

INDEX

Insulin, 10, 16
Intravenous injection, 66
Isoprenaline, 29
Isoproterenol, 29

Judgment, 36

Kidney, 33
Kosterlitz, Hans, 60

Lactose, 68
Lanterman-Petris-Short Act, 149
Large intestine, 33
Laws
  federal, 138–142
  state and local, 142–146
  treatment, 149
Lethal dosage, 47
Leukopenia, 72
Librax, 84
Librium, 84
Life span, effects of alcohol, 179–180
Limbic system, 36–37, 40, 42–47
Liver, 31
  effects of alcohol, 173
Local drug action, 21
Loss of control, 33
Loss of control, alcoholic, 189
LSD. See Lysergic acid
Lysergic acid, 93–96

Mainlining, 50
Mannite, 68
Marijuana, 1, 51, 57, 84–89
  laws, 141, 146
  pharmacological classification of, 51
Marijuana Tax Act, 138
Maslow, A. H., 126
Maximal dosage, 47
Medulla oblongata, 38–39
Medullary paralysis, 57, 59, 64
Menningar, Karl, 126–127
Meprobamate, 84
Mescaline, 91–92
Methadone, 153, 154
Methprylon, 79
Methylmorphine. See Codeine
Milk, 11, 33
Milk-sugar. See Lactose
Miltown, 76, 84
Minimal dosage, 47, 51
Monoamine, 30, 44
Monoamine oxidase (MAO), 46
Monoamine oxidase inhibitors (MAO inhibitors), 10–11, 13, 15–18, 100
Mood-modifiers, 30, 33
Moods, levels of, 44
Morning-glory seeds, 56
Morphine, 18, 60, 68–69
Motor, 34
Motor control, 40–42
Muscle coordination, 38
Muscles
  control of, 40–42
  effects of alcohol, 172

Nalline, 151, 153
Naloxone, 153
Names
  brand, 5
  chemical, 6
  common, 6
  generic, 5
  official, 5
  slang, 6
  street, 6
  trademark, 5

Narcolepsy, 104
Narcosis, 57
Narcotic Addict Rehabilitation Act, 141
Narcotic antagonists, 153–154
Narcotic Control Act, 139
Narcotic Drugs Import and Export Act, 138
Narcotics, 1, 12, 17, 59–65
  administration of, 65
  effects of, 59–62
  receptors of, 59
Narcotic-solvent abstinence syndrome, 62
Nerve cell-body, 43
Nervous system, 28
Neurohormone, 30, 44, 46
Neuromuscular junction, 28
Neuron, 28, 43
Neurotic state, 46
Neurotransmitter, 28, 43
New York Academy of Medicine, 154
Nitroglycerin, 40
Noludar, 79
Noradrenalin. See Norepinephrine
Norepinephrine, 29, 43, 46
Nowlis, Helen H., 146
Nutmeg, 56

Odyssey House, 155
Older problem drinkers, 187
On the nod, 62
Opheim, K. E., 59
Opium, 67
Opium Poppy Control Act, 139
Oral contraceptives, 14–15, 17–18
Oregono, 57
Overstimulation, 36, 47
Over-the-counter drugs, 56

"Package," 68
Pain, 38, 46
"Paper," 68
Paraldehyde, 77–78
Parasympathetic, 44
Parasympathomimetic drugs, 44
Parasympathetic nervous system, 40
Pars intermedia. See Pituitary gland
Parsley, 57
Party, obligations of host or hostess, 176
PCP. See Phencyclidine
Peace pill, 57
Penicillin, 56
Perceptional distortions, 33
Percodan, 68
Peripheral nervous system (PNS), 28
Personality, 34, 122–123
Pharmacopeia of the United States, 5
Phencyclidine (PCP), 57–58
Phenobarbitol, 13
Phenothiazines, 83
Phenylethylamines, 91–92
Phobex, 84
Phoenix House, 155
Pituitary gland, 60
Pituitary opioid peptide (POP), 59–60
Placidyl, 79
Pleasure, 46
POP. See Pituitary opioid peptide
Postganglionia, 43
Postganglionic neuron, 44
Patent medicines, 7
Preganglionic, 43
Preganglionic neuron, 44
Procaine, 68
Proof, 164
Proprietory, 6, 7
Proprioceptive sensations, 40
Psilocybin, 93

INDEX

Psychedelic, 90
Psychic energizers, 98–100
Psychoactive, 30–31, 33
Psychological effects, 46–47
Psychomotor activity, regulation of, 45
Psychosomatic effects, 46–47
Psychotic state, 46
Psychotogenic, 90
Psychotomimetic, 90
Psychotoxic, 30
Psychotropic, 30, 33
Pulse, 34
Punishment, 46
Pure Food and Drugs Act, 138

Rage, 46
Rashes, 46
Reasoning, 36
Receptor sites, 31
Recreational drugs, 56
Reflexes, 34
Reserpine, 83
Respiration, 34
Respiratory center, 38
Restless, 36
Reticular activating system, 37, 38, 44
Reticular formation, 36, 49–52
Reward, 46
Rockefeller University, 154

Saliva, 33
Scopolamine, 101
Scotch broom, 56
Sedation, 59
Sedatives, 13, 51
Senses, effects of alcohol, 172
Sensory, 34
Serotonin, 30, 45
Sexual effects of alcohol, 174
Side actions, 47
Sigeit, 34
Skeletal muscles, changes in, 46
Skin, effects of alcohol, 172
Skin popping, 50
Sleep, 34, 45
Sleep-eze, 81
Sleep-wake-mechanisms, 38
Smoking, 17–18
Sniffing, 69
Snorting, 69
Snyder, Solomon H., 60
Sobering up, 174
Social adjustment, 62
Sodium penthothal, 57
Sodium thiopental. See Sodium penthothal
Solvents, 52
Somnolence, 46
Spaced out, 58
Spectrum and Continuum of drug actions, 47–52
Speed cycle, 106
Spinal cord, 39
Step on, 68
Stimulant, 33
Stomach habit, 68
STP, 92
Stupor, 34
Suicide poison, 75
Sulfonamides, 10
Superego, 124

Sympathetic nervous system, 40
Sympathomimetic drugs, 44
Synanon, 125, 155–156, 195
Synapse, 31, 43
Systemic drug action, 21–24

Taste, 34
Temperature, 34
Teschemacher, H., 59
Tetracycline, 5
Thalamus, 37–38
Therapeutic communities, 154–156
Therapeutic dosage, 47
Therapy, 4
  chemo, 5
  dietary, 4
  operative, 4
  physical, 4
  psycho-, 4
Thinking, 36
Thinking disturbances, 33
Thin layer chromatography, 151–152
Thorazine, 99
Threshold ability, of pain and stress, 60
Thrombocytopenia, 72
Thyroid drugs, 17
Time or space, changes in, 33
TLC test, 152
Tolerance, 31, 62
Touch, 34
Toxic dosage, 47
Tranquilizers, 13, 82–84
  major, 82–84
  minor, 12, 83–84
Transmitter substance, 44
Treatment
  drug abuse, 148–157
  laws concerning, 149
Trophotropic system, 44–46
Tuberculin skin test, 18
Tyramine, 18
Tyrosine, 18

Ulcers, 46
Unconsciousness, 34
Untoward effects, 33, 47
Uppers, 56
Urine, 33
Users
  occasional, 116
  regular, 116

Valium, 15, 84
Valmid, 79
Vasomotor center, 38
Vitamin B_6, 18
Vitamin B_{12}, 10–11
Vitamin C, 14
Vitamin K, 11
Vomiting center, 39

Wakefulness, 38–39
Withdrawal, 51, 119, 152–153
Withdrawal therapy, 152
Women and alcoholism, 186–187
Wonder drugs, 2

X-ray, 15

Young people and alcohol, 177